T0134568

Marine Coastal Ecosystems Modelling and Conservation

Marco Ortiz • Ferenc Jordán
Editors

Marine Coastal Ecosystems Modelling and Conservation

Latin American Experiences

Editors
Marco Ortiz
Laboratorio de Modelamiento de Sistemas Ecológicos Complejos (LAMSEC)
Instituto Antofagasta
Universidad de Antofagasta
Antofagasta, Chile

Instituto de Ciencias Naturales Alexander von Humboldt, Facultad de Ciencias del Mar y Recursos Biológicos
Universidad de Antofagasta
Antofagasta, Chile

Departamento de Biología Marina
Facultad de Ciencias del Mar
Universidad Católica del Norte
Coquimbo, Chile

Ferenc Jordán
Balaton Limnological Institute
Centre for Ecological Research
Tihany, Hungary

Stazione Zoologica Anton Dohrn
Napoli, Italy

ISBN 978-3-030-58213-5 ISBN 978-3-030-58211-1 (eBook)
https://doi.org/10.1007/978-3-030-58211-1

This Springer imprint is published by the registered company Springer Nature Switzerland AG
The registered company address is: Gewerbestrasse 11, 6330 Cham, Switzerland

Foreword

When I was invited by the editors to write the foreword to this book, I became excited and agreed immediately. I read through the ten book chapters and my memories brought me back to the early years of the 1980s, when I started to get involved with fisheries modelling and teaching at different institutions of this continent and when I also learned about the tremendous role that the El Nino–Southern Oscillation (ENSO) plays in the marine ecosystems of Latin America.

Over the past four decades, I was able to witness the great progress in fisheries and ecological modelling on this continent and through my role as a researcher at the Centre for Tropical Marine Research (ZMT) and university professor at the University of Bremen, Germany, I supervised many master and doctoral students and participated in numerous research projects in Latin America related to fisheries modelling and conservation.

The book here presented addresses the challenges of reconciling fisheries management with ecosystem conservation and shows how difficult it is to accommodate the social and economic interests of resource users and the conservation of biodiversity and ecosystem health at the same time. The book presents a great variety of modelling approaches currently applied for supporting decisions in the fields of conservation, management and policy development. The chapters cover different ecosystems and resources, spanning from coral reef systems in the Caribbean Sea, kelp forest systems of Northern Chile, over demersal and pelagic systems of the Humboldt Current of Chile and Peru to the pelagic sardine fishery in the Gulf of California, México.

The book chapters nicely show how different modelling approaches allow integrating system relevant information and to holistically describe and quantify the threats and drivers that currently impact the marine systems and its resources. While fishing as a main driver of system change is shown by the different models to greatly impact the trophic structure and resilience of the ecosystems studied, other important drivers of system change are also considered, such as nutrient enrichment, deoxygenation and warming-induced increase in water column stratification, all of these potentially leading to changes in the nutrient supply, primary production, energy flow structure and key roles of species in the ecosystems. Several contributions demonstrate how climate-change induced spatio-temporal alterations in temperature, wind and current patterns can lead to critical conditions for the fishery resources exploited with great consequences for the associated fisheries.

The editors have managed to bring together renowned researchers and modellers from Latin America, each of them contributing to a better understanding of the functioning of the systems studied, thereby providing an improved basis for a more sustainable management and conservation of the systems studied. I like to congratulate the editors for this initiative and think that this book may represent a milestone on the way towards a better management and conservation of the marine systems and resources of Latin America.

Leibniz Centre for Tropical Marine Research (ZMT)
Bremen University, Germany

Matthias Wolff

Preface

Marine ecosystems are rich in biodiversity and bring richness to the ones overexploiting them. Especially in coastal zones, human impact is huge, ranging from overfishing to pollution. In order to better understand and preserve these beautiful ecological systems, we need information: data, models and research results. Management and conservation must be based on solid data and trustable information.

This book presents a small collection of excellent scientists making research on some aspects of coastal marine ecology, in one of the regions of Latin America. The aim is to bring together various challenges, approaches and modelling tools. We can see that some are specific, others are quite general. The authors of the chapters contributed to this book with great enthusiasm and we all agree that this book will contribute to smarter and more sustainable management of marine life.

There are many ways on how to model ecological systems and in this book, there is some emphasis on network models. The reason is that the complexity of coastal marine systems should be represented by integrating a large amount of information here. From phytoplankton to sharks and from zooplankton to jellyfish, the diversity of coexisting organisms must be synthesized somehow in order to give a full picture, in a holistic manner, to the stability, vulnerability or health of these ecosystems. All of the chapters contribute somehow to this system-based view on ecology.

Coquimbo, Chile — Marco Ortiz
Tihany, Hungary — Ferenc Jordán

Contents

Part I
Natural and Human Environment of Coastal Ecosystems

Ecological Modeling and Conservation on the Coasts of Mexico

L. E. Calderón-Aguilera, H. Pérez-España, R. A. Cabral-Tena, C. O. Norzagaray-López, A. López-Pérez, L. Alvarez-Filip, and H. Reyes-Bonilla

1 Introduction

The coasts of Mexico span for over 10,000 km in the Pacific and Gulf of Mexico and the Caribbean. Populated long before the Spaniards' arrival, the Mexican coasts have experienced drastic environmental changes during the last decades (Calderon-Aguilera et al., 2012).

This chapter aims to present an overview of the state of knowledge of selected Mexican coastal marine ecosystems' biodiversity, environmental condition, and anthropogenic impacts. Due to their biophysical and socio-economic importance,

L. E. Calderón-Aguilera (✉) · R. A. Cabral-Tena
Departamento de Ecología Marina, Centro de Investigación Científica y de Educación Superior de Ensenada, Ensenada, Baja California, Mexico
e-mail: leca@cicese.mx

H. Pérez-España
Instituto de Ciencias Marinas y Pesquerías, Universidad Veracruzana, Boca del Río, Veracruz, Mexico

C. O. Norzagaray-López
Instituto de Investigaciones Oceanológicas, Universidad Autónoma de Baja California, Ensenada, Baja California, Mexico

A. López-Pérez
Departamento de Hidrobiología, Universidad Autónoma Metropolitana – Iztapalapa, Ciudad de México, Mexico

L. Alvarez-Filip
Unidad Académica de Sistemas Arrecifales del Instituto de Ciencias del Mar y Limnología, UNAM en Puerto Morelos, Puerto Morelos, Quintana Roo, Mexico

H. Reyes-Bonilla
Departamento Académico de Ciencias Marinas y Costeras, Universidad Autónoma de Baja California Sur, La Paz, Baja California Sur, Mexico

M. Ortiz, F. Jordán (eds.), *Marine Coastal Ecosystems Modelling and Conservation*, https://doi.org/10.1007/978-3-030-58211-1_1

we focus on rocky reefs and coral reefs, taking into account their vulnerability to climate change.

1.1 Oceanographic Features

Flanked by the two big oceans, the coasts of Mexico are sharply different regarding their oceanographic characteristics. In the Atlantic, the coast is bathed by the Loop Current, which enters the Gulf of Mexico from the Caribbean, and the continental shelf is wide, predominately with soft bottoms. In the Pacific, the coast is narrow, waters are rich in nutrients and dissolved organic carbon, and it is under the influence of the California Current, the North-Equatorial Current, and the Costa Rica Current. Upwelling takes place in the Baja California Peninsula and the Isthmus of Tehuantepec. For their implications on modeling and conservation, we focus on two phenomes of increasing relevance: ocean acidification and shallowing of the minimum oxygen zone in the eastern Pacific.

Ocean Acidification

Approximately one-third of all the CO_2 released into the atmosphere since the industrial revolution has been absorbed by the oceans (Feely, Doney, & Cooley, 2009; Sabine et al., 2004). The absorbed CO_2 reacts with seawater, resulting in an increase in proton ions (H+), a reduction in pH, and a decrease in the levels of carbonate ion (CO_3^{2-}), consequently decreasing the saturation (Ω) of carbonate minerals (e.g., aragonite Ω_{Ar}) of seawater. This process, known as Ocean Acidification (OA), is projected to decrease the pH of oceanic surface waters by 0.14–0.35 pH units per the year 2100 (Hofmann et al., 2010; Orr et al., 2005). OA poses significant problems to marine organisms that form calcium carbonate skeletons, shells, or internal structures (Andersson, Mackenzie, & Bates, 2008; Cohen & Holcomb, 2009). OA research has been focused on its effect on the calcification of several organisms (Bhattacharya et al., 2016; Hoegh-Guldberg et al., 2007; Veron et al., 2009). Still, OA may affect other biological processes, such as reproduction (Kurihara, 2008), acid-base regulation (Portner, 2008), photosynthesis (Crawley, Kline, Dunn, Anthony, & Dove, 2010; Iglesias-Rodriguez et al., 2008), respiration (Rosa & Seibel, 2008), behavior (Munday et al., 2009), and tolerance to other stressors (Hoegh-Guldberg et al., 2007; Hutchins, Mulholland, & Fu, 2009).

Surface waters in many parts of the eastern tropical Pacific (ETP), including the Mexican Pacific (MP), have lower pH, lower Ω_{Ar}, and higher pCO_2 values in comparison to the rest of the tropics due to upwelling processes which mix CO_2 enriched deep waters into the surface layers (Manzello et al., 2008; Millero, 2007). This low pH condition has led to theorize that the ETP may give some insights into the biogeological interface in oceans with high levels of CO_2, and low Ω of carbonate minerals (Chapa-Balcorta et al., 2015; Manzello et al., 2008).

To date, most data regarding the variation of the pH, Ω_{Ar}, and the CO_2 system in the MP has been recorded by some oceanographic cruises (Franco et al., 2014; Hernández-Ayón, Zirino-Weiss, Delgadillo-Hinojosa, & Galindo-Bect, 2007) or calculated using data available in the World Ocean Atlas (Cabral-Tena et al., 2013; Manzello et al., 2008; Reyes-Bonilla, Calderón-Aguilera, Mozqueda-Torres, & Carriquiry-Beltrán, 2014; Saavedra-Sotelo et al., 2013) which may not necessarily represent the actual levels of variability along the coast or oceanic islands of the MP region. Nevertheless, two studies (Chapa-Balcorta et al., 2015; Norzagaray-López et al., 2017) described the seasonal variability and oceanographic processes that influence the CO_2 system in two areas of the MP.

Chapa-Balcorta et al. (2015) report the CO_2 system variables in the Gulf of Tehuantepec (GOT), located in the MP. Northerly winds intermittently influence the GOT, called "Tehuanos" (speed > 10 m s^{-1}) from November to March (Trasviña et al., 1995). These winds produce big changes in the structure of the water column and a complex coastal circulation. Chapa-Balcorta et al. (2015) described four oceanographic phenomena influencing the CO_2 system in the GOT: (1) A highly mixed region due to previous Tehuano events with surface Dissolved Inorganic Carbon (DIC) concentrations exceeding 2200 mmol kg^{-1} and pCO_2 levels above 1600 µatm, and very low Ω_{Ar} (1.2), (2) Upwelling areas observed in the coasts, which promote a DIC and pCO_2 enrichment, and a drop in pH and Ω_{Ar} in surface water near the coast, (3) Mesoscale eddies; the sinking of the thermocline by anticyclonic eddies contributed to surface water with minimum DIC and pCO_2 levels, pH and Ω_{Ar} maxima, (4) A poleward surface coastal current. This current functions as a DIC advection mechanism. The current runs through the mixed area and transports DIC-rich water to coastal regions located west of the GOT, thereby contributing to high DIC concentrations and low pH and Ω_{Ar} levels.

The low pH and Ω_{Ar} levels described in the GOT region (Chapa-Balcorta et al., 2015) may have adverse effects on the abundance, distribution, and calcification rate of corals in the area. However, coral communities in the area have been described as a mixture of patches along the coast (López-Pérez & Hernández-Ballesteros, 2004), with high coral cover (72.78 ± 2.4%), species richness (14 spp) and with similar calcification rates of branching (*Pocillopora* spp: 2.99–5.12 $gCaCO_3$ cm^{-2} yr^{-1}) and massive (*Pavona* and *Porites* spp: 0.34–0.45 $gCaCO_3$ m^{-2} yr^{-1}) species in comparison to other MP and ETP areas (López-Pérez et al., 2014; López-Pérez & López-López, 2016; Medellín-Maldonado et al., 2016). The latter can be explained because the coastal zone displays high variability and is also influenced by other processes (river inputs, eutrophication, waves) that modulate the degree of exposure of corals to corrosive water (Chapa-Balcorta et al., 2015).

Norzagaray-López et al. (2017) described the seasonal variability of the CO_2 System in Cabo Pulmo (CP), a fringing coral reef in the entrance to the Gulf of California. The authors described the influence of two sea surface water masses: Gulf of California Water (GCW) during the winter, with low temperature (21.6 ± 2.1 °C), pH (7.93 ± 0.02), and Ω_{Ar} (2.7 ± 0.2), and high DIC levels (2079 ± 27 µmol kg^{-1}); and Tropical Surface Water (TSW) during the rest of the year, with higher temperature (26.9 ± 2.9 °C), pH (8.01 ± 0.3), and Ω_{Ar} (3.2 ± 0.3), and lower DIC levels (2008 ± 48 27 μmol kg^{-1}). The seasonal variability in water masses in the GC results from forcing events (Lavín & Marinone, 2003). The CO_2 system at CP reef stems from a combination of (1) the ridge-like geomorphology, which facilitates water circulation (Riegl, Halfar, Purkis, & Godinez-Orta, 2007); (2) the direct communication with the adjacent ocean; and (3) the seasonal changes in the oceanographic conditions in the area. The CP reef experiences a seasonal variability in Ω_{Ar}, with a difference of 0.5 units, depending on which water mass is present (TSW or GCW), which is one of the largest ranges recorded in reefs worldwide (Manzello et al., 2008; Manzello, Enochs, Melo, Gledhill, & Johns, 2012). Seasonal fluctuations in the CO_2 parameters in reef areas have been explained by several mechanisms related to the reef geomorphology and circulation patterns, which define the exchange of chemical properties between the reef and the adjacent sea by regulating the flow of seawater and its residence time (Andersson, Yeakel, Bates, & de Putron, 2014; Manzello et al., 2008, 2014; Muehllehner, Langdon, Venti, & Kadko, 2016; Rixen, Jiménez, & Cortés, 2012; Shaw, McNeil, & Tilbrook, 2012). In exposed reef environments, oceanic processes (such as horizontal advection or upwelling) might be the most critical controls. Particularly, DIC and temperature analyses confirmed that Cabo Pulmo has direct communication with open seawater (Norzagaray-López et al., 2017), which implies that if the benthic reef community modifies the carbonate system, this signal could be masked by ocean water, thus making the ability of the benthic community to buffer or modify the carbonate system very limited (Andersson et al., 2014; Manzello et al., 2012; Muehllehner et al., 2016; Shaw et al., 2012).

CP is one of the most well-preserved coral communities in the ETP, and among the most relevant sites for conservation in the western coast of Mexico (Aburto-Oropeza et al., 2011; Enochs & Glynn, 2017; Reyes-Bonilla et al., 2014) which has resulted, among other things, in a huge fish biomass and diversity (Ayala-Bocos, Fernández-Rivera-Melo, & Reyes-Bonilla, 2018; Reyes-Bonilla, Álvarez del Castillo-Cárdenas, et al., 2014). The scleractinian coral community of Cabo Pulmo is composed of 12 species and is dominated by *Pocillopora* (Glynn et al., 2017; Reyes-Bonilla, 2003; Reyes-Bonilla, Sinsel-Duarte, & Arizpe-Covarrubias, 1997). Coral cover in Cabo Pulmo is approximately 18–27% (Norzagaray-López et al., 2017; Reyes-Bonilla & Calderon-Aguilera, 1999) and has an estimated net carbonate production of 3.48 kg m^{-2} yr^{-1} (Cabral-Tena et al., 2018). Nevertheless, Cabo Pulmo has a narrow reef framework (Calderon-Aguilera, Reyes-Bonilla, & Carriquiry-Beltran, 2007; Glynn et al., 2017). Most of the reef derived sediments are transported off-reef (Riegl et al., 2007), but, more importantly, because of the described seasonal variations in the Ω_{Ar} of 0.5 by Norzagaray-López et al. (2017). Thus, an intensification of the seasonal variations in the Ω_{Ar} in Cabo Pulmo could lead to potentially large-scale impacts that are likely to affect the ecosystem's metabolic processes (Andersson et al., 2014; Veron et al., 2009).

Cabo Pulmo has been the focus of many studies, including modeling; Reyes-Bonilla, Álvarez del Castillo-Cárdenas, et al. (2014) through a mass-balance model estimated biomass export around 5330 t/km^2/yr, worth USD 177,100 per year. Network analysis by Calderon-Aguilera and Reyes-Bonilla (2016) showed that despite being protected for 20 years, the ecosystem is not mature yet.

Oxygen Minimum Zones

The ETP is one of the most extensive and intense oxygen minimum zones (OMZ) in the world (Karstensen, Stramma, & Visbeck, 2008; Keeling, Kortzinger, & Gruber, 2010; Stramma, Johnson, Sprintall, & Mohrholz, 2008). These hypoxic conditions are present due to an intense organic matter production in the photic zone, which is consequently remineralized during falling in the water column, mainly by oxic respiration processes by heterotrophic bacteria groups (Ulloa, Canfield, DeLong, Letelier, & Stewart, 2012). These conditions are maintained by poor ventilation resulted from a strong stratification (Karstensen et al., 2008). Dissolved oxygen concentration is an important factor for many biogeochemical processes occurring in the water column, for instance, low oxygen concentrations promote chemical speciation of several elements, controlling their availability or low oxygen concentrations define the habitat for many macro-organisms (Keeling et al., 2010; Stramma et al., 2008; Vaquer-Sunyer & Duarte, 2008). Thresholds on oxygen concentrations to define hypoxia levels are still in debate, however, when oxygen content reaches suboxic conditions ($O_2 < 20$ μmol kg^{-1} or about 10% of sea surface values) they are considered as seriously hypoxic conditions for several taxa (Gilly, Beman, Litvin, & Robinson, 2013; Keeling et al., 2010; Paulmier, Ruiz-Pino, & Garcon, 2008; Stramma et al., 2008). This oxygen value is close to the anaerobic microbial processes limit (5–10 μmol kg^{-1}). In addition, the intense remineralization of the organic matter in OMZs modifies the cycles of other elements (Fe, P, N, C); with regard to the carbon cycle, these zones are also known as maximum carbon zone, with high levels of dissolved inorganic carbon and low pH values (Paulmier et al., 2008).

In the tropical Pacific off Mexico converge three oceanographic features (Cepeda-Morales, Beier, Gaxiola-Castro, Lavín, & Godínez, 2009; Franco et al., 2014; Prince & Goodyear, 2006): (1) the OMZ located below a shallow thermocline, and a depth between 70 and 110 m, controlled by mesoscale physical processes (i.e., coastal upwelling, advection, eddies), (2) a shallow maximum carbon zone, and (3) the upper limit of the subsurface subtropical water, a cold, nutrient-rich and high inorganic content water mass. These oceanographic conditions influence the ecosystems developing in shelf and coastal environments, both benthic and pelagic (Prince & Goodyear, 2006; Vaquer-Sunyer & Duarte, 2008). The biological implications of the OMZ at taxa level are highly variable because each group has developed a behavioral and/or physiological aspect to cope with hypoxic conditions (Prince & Goodyear, 2006; Vaquer-Sunyer & Duarte, 2008). Some of these organisms are important for commercial fisheries and entertainment activities. Vaquer-Sunyer and Duarte (2008) reported a broad variability in the physiological response in benthic organisms (i.e., lethal concentration and exposure time), finding that crustacean and fishes are the more sensitive groups. In contrast, the most tolerant group was mollusks. This benthic fish and crustacean species use several mechanisms to cope with suboxic conditions; for example, they reduce their metabolic rate in order to reduce the oxygen demand. In the case of the shallow OMZ from the ETP, this water is corrosive and hypoxic, and both conditions are predicted to have a detrimental effect on the physiology of calcifying organisms (low temperature, low pH, and reduced saturation state regarding carbonate mineral phases).

The OMZ is relevant for pelagic ecosystems because the upper boundary layer plays an essential role as a refuge for many zooplankton taxa (krill, myctophid fishes) from predators during the daytime (Gilly et al., 2013). During the night, this micronekton migrates to shallower layers to feed on phytoplankton. This behavior causes zones of dense aggregations of organisms, which are foraging for many deep-dive pelagic predators, macro-organisms like swordfish, squid, and tuna. Thus, these conditions modify the behavior of pelagic fish species, defining the habitat use in the water column (e.g., tuna, sailfish, billfish). These species are particularly sensitive to hypoxic conditions due to their high metabolic rate (Brill, 1996). For example, the sailfish and billfish spend most of the time in shallower zones, suggesting that both temperature and oxygen concentration, are the main factors defining their habitat preference (Prince & Goodyear, 2006). This behavior also indicates that these species prefer the mixed layer (above the thermocline), a more oxygenated zone with the higher primary production driven by a nutrient and light availability. Thus, the OMZ from ETP is a link between mesopelagic and epipelagic environments in a series of trophic networks (Robison, 2009), and the expansion and shallowing of the OMZs will have a profound effect on it. The OMZ from the ETP has maintained suboxic conditions and expanded the vertical distribution in the water column (shallower) during the last four decades (Stramma et al., 2008). This OMZ shallowing agrees with climatic models that highlight a decrease in the oxygen concentration and an expansion of the OMZs, promoted by a stronger stratification driven by rising temperatures, thus reducing the ventilation of the water column (Keeling et al., 2010). Also, this layer contributes to greenhouse gases (CO_2 and N_2O). As a result of anthropogenic activities, increased stratification and oxygen depletion must be considered with the same concern that rising temperatures and ocean acidification predictions (Portner & Farrell, 2008).

1.2 Biodiversity of Coral and Rocky Reefs

Shallow coral reefs in Mexico occupy some 1068.289 km^2 (Caribbean, 672.8615, Burke, Reytar, Spalding, & Perry, 2011; Gulf of Mexico, 384.83, Tunnell, 2007; Mexican Pacific, 10.597, Lopez-Perez et al. in press, Reyes-Bonilla and Lopez-Perez unpublished data), and only 0.032% of the exclusive economic zone of

Mexico. Though scarce in extension, it is a critical resource since they supply vast numbers of people with various goods and ecological services (Moberg & Folke, 1999). Goods and ecological services are generated and sustained by biological communities, and these, in consequence, sustaining ecosystem health and resilience (Duffy et al., 2013). Reef degradation and biodiversity loss may impair the reef's capacity to provide goods and services and recover from perturbations (Hughes et al., 2017; Moberg & Folke, 1999; Worm et al., 2006), therefore managing reef systems to maintain its biodiversity may provide a way to guarantee their permanence across space and time. Nevertheless, our knowledge of reef-associated biodiversity in Mexico is still scarce.

There are few extensive compilations of marine biodiversity associated with coral systems in Mexico; most of the accounts are partial compilations for some taxonomic groups, partial taxonomic list for particular areas, or large-scale compilations including all coastal and marine ecosystems (Table 1). What follows is a brief biodiversity account by oceanic basin.

Table 1 Number of reported species of selected taxonomic groups from the coasts of Mexico

	Caribbean[3,4]	Gulf of Mexico	Pacific
Algae	561	493[1,6]	177[4,7,43]
Sponges	519	93[2,8]	87[43,44]
Cnidarians	994	104[10,11,12,14,23]	157[2,5,9,10,36,37,43,44]
Annelids	658	41[13]	329[22,44]
Mollusks	3032	132[20,27,30]	297[8,35,39,40,43,44]
Arthropods	2916	237[15,17,18,19,21,25,26,29]	418[6,31,32,33,38,41,43,44]
Echinoderms	438	51[24]	75[42,43,44]
Fishes	1336	312[22]	372[1,3,43]
Total	10,454	1463	1912

1. Thomson et al. (2000), 2. Reyes-Bonilla et al. (2005), 3. Alvarez-Filip, Reyes-Bonilla, and Calderon-Aguilera (2006), 4. Pedroche et al. (2005), 5. Breedy & Guzmán (2007), 6. Martínez-Guerrero (2007), 7. Pedroche et al. (2008), 8, Ríos-Jara et al. (2008), 9. Breedy, Guzman, and Vargas (2009), 10. Cairns and Fautin (2009), 11. Cairns, Jaap, and Lang (2009), 12. Calder and Cairns (2009), 13. Fauchald, Granados-Barba, and Solís-Weiss (2009), 14. Fautin and Daly (2009), 15. Felder, Álvarez, Goy, and Lemaitre (2009), 16. Fredericq et al. (2009), 17. Gittings (2009), 18. Heard and Anderson (2009), 19. Heard and Roccatagliata (2009), 20. Judkins, Vecchione, and Roper (2009), 21. LeCroy, Gasca, Winfield, Ortiz, and Escobar-Briones (2009), 22. McEachran (2009), 23. Opresko (2009), 24. Pawson, Vance, Messing, Solís-Marin, and Mah (2009), 25. Price and Heard (2009), 26. Reaka et al. (2009), 27. Rosenberg, Moretzsohn, and García (2009), 28. Rützler, van Soest, and Piantoni (2009), 29. Schotte, Markham, and Wilson (2009), 30. Turgeon, Lyons, Mikkelsen, Rosenberg, and Moretzsohn (2009), 31. García-Madrigal & Andreu-Sánchez (2009), 32. Hernández et al. (2009), 33. Hernández et al. (2010), 34, Miloslavich et al. (2010), 35. Reyes-Gómez (2016), 36. Breedy and Guzman (2011), 37. Abeytia, Guzmán, and Breedy (2013), 38. Hernández et al. (2013), 39. Flores-Rodríguez et al. (2014), 40. Barrientos-Luján et al. (2017), 41. Valencia-Mendez, Lopez-Perez, Martinez-Guerrero, Antonio-Perez, and Ramirez-Chavez (2017), 42. Aparicio-Cid (2018), 43. Lopez-Perez et al. (in press), 44. Gulf of California database (2019, https://www.desertmuseum.org/center/seaofcortez/database.php)

Caribbean

The Caribbean Sea is considered a unique biogeographic region, constitutes 12% of the total reefs of the world, and is considered a global-scale hot spot of marine biodiversity (Roberts et al., 2002). However, the Caribbean has a complex geological history, and the present-day geographic diversity in hydrologic, morphologic, and habitat regimes has led to the recognition of several distinct biogeographic sectors with relatively homogenous species composition (Spalding et al., 2007). Although, the limits of some of these biogeographic sectors are likely to vary depending on the taxa of interest, it is generally accepted that one of the most clearly define eco-regions within the Caribbean is the Western Caribbean, or as it is commonly known: the Mesoamerican Reef (MAR). The Mexican Caribbean is part of this eco-region that extends over 1000 km along the state of Quintana Roo, Mexico, Belize, and the Caribbean coasts of Guatemala and Honduras (Rioja-Nieto, Garza-Pérez, Álvarez-Filip, Ismael, & Cecilia, 2019). The coral reef systems of the Mexican Caribbean extend over 400 km of coast and comprise an inter-mix of coral reefs, seagrasses, algal beds, and areas of soft substrate, inextricably connected in terms of species movement and energy flow, interacting to persist under natural and anthropogenic disturbances, and climate change (Rioja-Nieto et al., 2019).

In the Mexican Caribbean, biodiversity studies have primarily focused on scleractinian corals and reef fishes, two diverse and functionally important groups on reef environments (Table 1). Coral reef research efforts in the region started in the late 1970s with a through region-wide description of coral communities across reef-zones up to 30-m depth (Jordán-Dahlgren, 1993), since then several efforts have been made to describe biogeographic patterns, key ecological drivers of biodiversity, ecological interactions and assess the effect of disturbances and management strategies (Alvarez-Filip, Côté, Gill, Watkinson, & Dulvy, 2011; Rioja-Nieto & Álvarez-Filip, 2019).

Studies exploring biodiversity patterns across have found that reef attributes such as reef zone, reef size, or reef complexity are better descriptors of diversity patterns than latitudinal gradients. For example, Arias-González, Legendre, and Rodríguez-Zaragoza (2008) studied variation in fish and coral species diversity at two spatial scales: among geomorphology classes (reef lagoons, fronts, slopes, and terraces) within reefs (beta diversity), and across a 400-km latitudinal diversity gradient (delta diversity); finding that the geomorphological structure, "reefscape" attributes at different scales, and depth are important variables for shaping beta/delta diversity of fish and coral species richness and community composition at the various scales. Rodríguez-Zaragoza and Arias-Gonzalez (2008), using fish communities to explore diversity partitioning within the Mexican Caribbean, found that total fish diversity is determined mostly by reef scale followed per sample and habitat scales; which supports the hypothesis that inter-habitat and reef differences seem to regulate local and regional species richness strongly. Other studies have also found that species richness of corals and fish are strongly correlated with reef attributes such as reef area and habitat complexity (rather than geographical location) and were the main proxies for critical coral reef biodiversity values (Rodríguez-Zaragoza &

Arias-Gonzalez, 2015). In recent work, González-Barrios and Álvarez-Filip (2018) proposed a Reef Functional Index (RFI) that considers the calcification rate, structural complexity, and cover of each coral species. They claim that this index can be used to compare changes in coral community composition, keystone species loss, and functional loss due to climate change and other stressors.

The effects of rapid ecological degradation and human-related threats on diversity patterns of reef communities have also been assessed recently. Acosta-Gonzalez, Rodríguez-Zaragoza, Hernández-Landa, and Arias-Gonzalez (2013) showed losses of α, β, and γ-diversity of fish communities associated with temporal coral-algal phase shifts in the region. Despite a drastic reduction in the number of species over time, β-diversity continued to be the highest component of γ-diversity. The shift transition harmed α, β, and γ-diversity, primarily by impacting rare species, leading a group of small and less vulnerable fish species to become more frequent and another group of rare species to become locally extinct. The maintenance of fish heterogeneity (β-diversity) over time may imply the abetment of vulnerability in the face of local and global changes (Acosta-Gonzalez et al., 2013). In the most extensive temporal study published for the region, Schmitter-Soto et al. (2018) provide evidence of the rapid ecological changes reef fishes underwent between 1994 and 2015 in the southern Mexican Caribbean (Mahahual to Xcalak). The mean density of most species and guilds decreased significantly through time. Fish species from higher trophic levels (e.g., piscivores) were among the most affected.

Gulf of Mexico

Gulf of Mexico is a basin located in the southeast of the USA, northeast of México, and west Cuba; it comprises an area of 564,200 km^2 of which, 35.2% is underlain by continental shelf (Darnell & Defenbaugh, 1990). Coral reefs growths mainly on the shelf of Veracruz, Campeche, and the Yucatán States at the Mexican side, as well as on the West and South coast of Florida shelf in the USA. Also, there are isolated growth formations at the outer shelf of Texas. In General, while in the Caribbean, there are favorable conditions to develop an extensive, almost continues barrier formed by corals; in the Gulf of Mexico, coral reefs are present only in areas where the conditions of the substrate and oceanographic variables are appropriate (Jordán-Dahlgren & Rodríguez-Martínez, 2003).

North and West shelves of the Gulf of Mexico are not an adequate place for the growth of coral reefs, as it is mainly a terrigenous area, under the influence of large rivers like the Mississippi or Grijalva. When Heilprin (1890) visited Veracruz, southwest Gulf of Mexico, he was surprised by the existence of corals forming abundant reefs. Despite being in a place with a long story of scientific studies, the Mexican part of the Gulf of Mexico is still not wholly unstudied. During 2017, 17 "new" reefs were discovered in the MPA Parque Nacional Sistema Arrecifal Veracruzano (Camarena-Luhrs, Gray-Vallejo, Liaño-Carrera, & Aragón-González, 2017).

Regarding the biodiversity in coral reefs (Table 1), there is a tendency to decrease from the Caribbean to the West and the North; it happens with gorgonians (Jordán-Dahlgren, 2002) and corals (Horta-Puga, Vargas-Hernández, & Carricart-Ganivet, 2007), but possibly also with fishes (Robertson & Cramer, 2014) and another poorly studied taxonomic groups. The Gulf of Mexico has a relatively high number of endemic species. Robertson and Cramer (2014) identified that 9% of fish species are endemic, and groups like sponges have up to 32% of endemic species (Rützler et al., 2009).

About models in the Gulf of Mexico, there are some oceanographic models as well as mass-balance models for coral reefs. Salas-Monreal et al. (2018) studied the connectivity of coral reefs within the western Gulf of México. They found high connectivity through two pathways: Campeche Bank and Veracruz and Tuxpan reefs and a second pathway between Tuxpan and Flower Garden Banks. Though there are some mass-balance models built with Ecopath since 1993 in the Gulf of Mexico (e.g. Arreguín-Sánchez, Seijo, & Valero-Pacheco, 1993), there are few or none published models specific for coral reefs.

Pacific

In the Mexican Pacific, corals and coral reef systems extend from the head of the Gulf of California (Reyes-Bonilla & López-Pérez, 2009) to the Gulf of Tehuantepec, Oaxaca (Glynn & Morales, 1997). Currently, the Gulf of California contributed 36% of the entire Mexican Pacific coral records, while the rest is located in oceanic islands such as Revillagigedo (Ketchum & Reyes-Bonilla, 2001) and along the Mexican mainland from Nayarit to Oaxaca (Reyes-Bonilla et al., 2005). Although geographically extended, Pacific systems encompass less than 1% of the total reef extension of the country.

In the Mexican Pacific most of the reef-associated biodiversity studies are local in scope (i.e., Cabo Pulmo), restricted to well-studied macrofauna such as corals (Reyes-Bonilla et al., 2005), echinoderms (Granja-Fernández et al., 2014), and fishes (Alvarez-Filip et al., 2006), incipient studies in some groups (Humara-Gil & Cruz-Gomez, 2018; Jarquín-González & García-Madrigal, 2010), but fully absent in other taxa. Additionally, early biodiversity accounts obviated the substrate on which the organism was collected and hence, still uncertain its association with reef systems.

As expected, the Caribbean hosts the most considerable species richness followed by the Mexican Pacific and the Gulf of Mexico (Table 1). While the trend is in agreement with large-scale spatial studies (Costello et al., 2010; Miloslavich et al., 2010), numbers should be considered with caution since they may not necessarily correspond to reef systems alone (i.e., the Caribbean, Gulf of California) or being restricted to the Mexican exclusive economic zone (i.e., Caribbean). Additionally, it is expected that future sampling of not yet surveyed habitats, sites, and taxa employing traditional and novel techniques may render far more species than the currently recognized.

Finally, biodiversity should be seen as a master variable for practically evaluating both the health of ecosystems and the success of management efforts. Hence, a standardized and systematic monitoring protocol under a marine biodiversity observation network is necessary for effective reef management.

1.3 Ecological Connectivity

Connectivity is a fundamental ecological process in marine ecosystems that promotes both the persistence and recovery of populations through the dispersal of marine life across populations, communities, and ecosystems (Pineda, Hare, & Sponaugle, 2007). Yet, it has been scarcely used in Mexican waters as a tool to address connectivity at any biological organization level or ecosystems, including reefs. A recent review, for example, pointed out that research effort on population connectivity in the design of marine protected areas is particularly low in the country, though as large as the effort conducted in Canadian waters (Balbar & Metaxas, 2019).

In Mexican waters, different methods have been used to model the connectivity of reef organisms. They include gene flow of populations of different taxa such as sponges (León-Pech, Cruz-Barraza, Carballo, Calderon-Aguilera, & Rocha-Olivares, 2015), corals (Martínez-Castillo, Reyes-Bonilla, & Rocha-Olivares, 2018; Saavedra-Sotelo et al., 2013), sea cucumbers (Ochoa-Chávez, Del Río-Portilla, Calderón-Aguilera, & Rocha-Olivares, 2018), and fish (Villegas-Sánchez, Pérez-España, Rivera-Madrid, Salas-Monreal, & Arias-González, 2014); biophysical modeling (Lequeux, Ahumada-Sempoal, López-Pérez, & Reyes-Hernández, 2018; Marinone, Ulloa, Parés-Sierra, Lavin, & Cudney-Bueno, 2008), and connectivity via multidisciplinary approaches (Munguia-Vega et al., 2014). In addition, other studies encompass several spatial scales ranging from relatively local in scope as the conducted in selected areas of the Gulf of California (Martínez-Castillo et al., 2018; Peguero-Icaza, Sánchez-Velasco, Lavín, Marinone, & Beier, 2011), and the Central Mexican Pacific (López-Pérez et al., 2015), to regional (Jordán-Dahlgren, 2002; Lequeux et al., 2018; Murphy & Hurlburt, 1999; Sanvicente-Añorve, Zavala-Hidalgo, Allende-Arandía, & Hermoso-Salazar, 2014), and as a part of global-scale studies (Wood, Paris, Ridgwell, & Hendy, 2014). Overall, while empirical methods are relatively expensive and require intensive sampling, biophysical modeling offers the possibility to study connectivity by means of tracking numerous virtual larvae over a wide range of spatio-temporal scales and under several scenarios, and hence have been commonly employed.

At mesoscale, overall connectivity patterns follow main current systems and its seasonal variations, as stated for a semi-enclosed sea such as the Gulf of California (Montaño-Cortés, Marinone, & Valenzuela, 2017; Munguia-Vega et al., 2014; Peguero-Icaza et al., 2011; Soria et al., 2014). It also holds for large areas such as the eastern Pacific (Lequeux et al., 2018; Romero-Torres, Treml, Acosta, & Paz-García, 2018) or the Gulf of Mexico/Caribbean sea (Murphy & Hurlburt, 1999;

Sanvicente-Añorve et al., 2014). As a consequence, commonly the number of particles is high across or close to the diagonal of the connectivity matrix, indicating that high self- and subsidiary recruitment between relatively close areas is the predominant processes unrestricted of the spatial and temporal scale scenario addressed (Garcés-Rodríguez et al., 2018; Johnston & Bernard, 2017; Lequeux et al., 2018). Concurrently, connectivity matrices also reveal that the larval exchange among nearby areas may potentially involve up to several thousand larvae per year. Hence, it may have potential effects on demographic connectivity (Lequeux et al., 2018). Still, as dispersal distances increased, larvae remain longer in the water column, subtle genetic connectivity as suggested by molecular data (Chávez-Romo et al., 2009; León-Pech et al., 2015; Paz-García et al., 2009, 2012; Saavedra-Sotelo et al., 2011, 2013; Villegas-Sánchez et al., 2014).

1.4 Main Threats to Biodiversity from Human-Related Activities in the Caribbean

In the past few decades, multiple threats have contributed to substantial declines in coral cover worldwide (Hughes et al., 2018). In the Caribbean, coral decline on many reefs has been associated with an increase in macroalgal cover (Jackson, Donovan, Cramer, & Lam, 2014). A combination of multiple factors, from climate change to the disruption of reef ecological coherence, drives this shift (Jackson et al., 2014). However, anthropogenic activities such as eutrophication, sedimentation, and local contamination of coastal waters are increasingly being associated to reef degradation at numerous sites globally (Baker, Rodriguez-Martinez, & Fogel, 2013). For example, unsustainable agriculture and coastal development can lead to elevated sediment and nutrient run-off, with detrimental impacts on nearby coral reefs, irrespective of marine protection (Bégin et al., 2016; Wenger et al., 2016). Sedimentation raises water turbidity, reducing coral photosynthetic activity, energy reserves and growth, and has been linked with elevated coral disease prevalence (Pollock et al., 2014); while excessive nutrients generated inland (e.g., sewage, fertilizers) reaching reefs are linked with the increase in the prevalence and severity of coral diseases and promote the exponential growth of macroalgae, which compete with corals (Suchley, McField, & Alvarez-Filip, 2016; Vega Thurber et al., 2014). Land-based threats can also interact synergistically with other stressors, for example, nutrient enrichment may increase the susceptibility of corals to climate-related coral bleaching (Wiedenmann et al., 2013). The Caribbean coast of Mexico has experienced dramatic coastal development over the last 30–40 years (Baker et al., 2013). Over 10 million tourists visit the region annually, and the local population has multiplied from 88,000 in 1970 to 1.5 million in 2015 (Suchley & Alvarez-Filip, 2018). Therefore, there are several concerns regarding anthropogenic impacts on the coastal ecosystems of the Mexican Caribbean (Rioja-Nieto et al., 2019; Suchley & Alvarez-Filip, 2018). These include loss of both forest and mangrove vegetation owing to hotel construction and urbanization, waste generation, overfishing, and groundwater pollution. The later represents the most severe threat to the reef system, as the coastal aquifer is highly vulnerable to contamination and discharges directly into the sea (Rioja-Nieto et al., 2019). Pollutants detected in groundwater discharges include heavy metals, hydrocarbons, chlorophenoxy herbicides, and fecal indicator bacteria (reviewed by Rioja-Nieto et al., 2019).

Land-based activities and in particular marine eutrophication resulting from inadequate wastewater treatment is considered a principal driver of declining reef condition in the Mexican Caribbean (Suchley et al., 2016). The high connectivity between these local-scale threats and coral reef condition has been established for the region. Baker et al. (2013) showed, using stable isotopes from gorgonians, that water enrichment (nitrogen source) is positively correlated with tourist visitation over a timescale of 7 years. On a local scale in the South Mexican Caribbean, Arias-González et al. (2017) reported a rapid phase shift from a coral-dominated system to a fleshy macroalgae-dominated system in a time of only 14 years. This phase shift can be primarily attributed to coastal landscape transformation induced by the development of a touristic infrastructure (hotel and restaurants) and cruise ship ports (Martínez-Rendis et al., 2015). In a larger spatial scale, Suchley and Alvarez-Filip (2018) found that the coral cover was significantly lower at sites with elevated local human activity. This study forecasted that if high rates of coastal development continue, then highly degraded coral reef sites with low coral cover are likely to become increasingly common. However, integrated coastal zone management, particularly if combined with a region-wide ban on herbivorous fish extraction, could mitigate the negative impacts of planned developments and improve benthic conditions beyond current levels (Suchley & Alvarez-Filip, 2018).

2 Conclusion and Remarks

Coastal ecosystems are intrinsically highly dynamic and affected by many environmental and anthropogenic factors. There is no feasible way to track those changes; therefore, a modeling approach is the most suitable way to identify, prevent, and, if possible, mitigate those impacts. Hereby we have presented some critical issues about both coasts of Mexico, and some of the modeling approaches followed. Considering the information available for the region, some of the models that will be used are:

(i) Mass-balance models. Based on the worldwide used software *Ecopath with Ecosim*, key questions as evaluating ecosystem effects of fishing, management options, and analysis of impact and placement of marine protected areas can be addressed with this modeling approach.
(ii) Ecological niche models. Under a climate change scenario, species distribution shifting is a primary concern. So ecological niche modeling would be conducted more often. The Maximum Entropy algorithm (MAXENT) has already been used in Mexico.

Network analysis. Closely related to mass-balance models, network analysis focuses on other ecosystem's properties such as *Throughput, Ascendancy*, and *Overhead*. Its use and application will increase as the input data become available.

(ii) Generalized Linear Models, including mixed, additive, and the kind. The use of coupled biophysical models, connectivity analysis, and many other statistical models is already standard practice, although many of those models are graduate thesis and have not been published yet.

References

Abeytia, R., Guzmán, H. M., & Breedy, O. (2013). Species composition and bathymetric distribution of gorgonians (Anthozoa: Octocorallia) on the Southern Mexican Pacific coast. *Revista de Biología Tropical, 61*(3), 1157–1166.

Aburto-Oropeza, O., Erisman, B., Galland, G. R., Mascareñas-Osorio, I., Sala, E., & Ezcurra, E. (2011). Large recovery of fish biomass in a no-take marine reserve. *PLoS One, 6*, e23601. https://doi.org/10.1371/journal.pone.0023601.

Acosta-Gonzalez, G., Rodríguez-Zaragoza, F. A., Hernández-Landa, R. C., & Arias-Gonzalez, J. E. (2013). Additive diversity partitioning of fish in a Caribbean coral reef undergoing shift transition. *PLoS One, 8*(6), e65665.

Alvarez-Filip, L., Côté, I. M., Gill, J. A., Watkinson, A. R., & Dulvy, N. K. (2011). Region-wide temporal and spatial variation in Caribbean reef architecture: is coral cover the whole story? *Global Change Biology, 17*(7), 2470–2477.

Alvarez-Filip, L., Reyes-Bonilla, H., & Calderon-Aguilera, L. E. (2006). Community structure of fishes in Cabo Pulmo Reef, Gulf of California. *Marine Ecology, 27*, 253–262.

Andersson, A. J., Mackenzie, F. T., & Bates, N. R. (2008). Life on the margin: implications of ocean acidification on Mg-calcite, high latitude and cold-water marine calcifiers. *Marine Ecology Progress Series, 373*, 265–273.

Andersson, A. J., Yeakel, L., Bates, N. R., & de Putron, S. J. (2014). Partial offsets in ocean acidification from changing coral reef biogeochemistry. *Nature Climate Change, 4*, 56–61. https://doi.org/10.1038/nclimate2050.

Aparicio-Cid C. (2018). Biogeografía de los equinodermos arrecifales del Pacífico mexicano. MSc Thesis. Centro Interdisciplinario de Ciencias Marinas, La Paz, Baja California Sur, México. 123 p.

Arias-González, J. E., Fung, T., Seymour, R. M., Garza-Pérez, J. R., Acosta-González, G., Bozec, Y. M., et al. (2017). A coral-algal phase shift in Mesoamerica not driven by changes in herbivorous fish abundance. *PLoS One, 12*(4), e0174855.

Arias-González, J. E., Legendre, P., & Rodríguez-Zaragoza, F. A. (2008). Scaling up beta diversity on Caribbean coral reefs. *Journal of Experimental Marine Biology and Ecology, 366*(1–2), 28–36.

Arreguín-Sánchez, F., Seijo, J. C., & Valero-Pacheco, E. (1993). An application of ECOPATH II model to the north continental shelf ecosystem of Yucatan (pp. 269–278).

Ayala-Bocos, A., Fernández-Rivera-Melo, F., & Reyes-Bonilla, H. (2018). Listado actualizado de peces del arrecife de Cabo Pulmo, Golfo de California, México. *Revista Ciencias Marinas y Costeras, 10*(1), 9–29.

Baker, D. M., Rodriguez-Martinez, R. E., & Fogel, M. L. (2013). Tourism's nitrogen footprint on a Mesoamerican coral reef. *Coral Reefs, 32*(3), 691–699.

Barrientos-Lujan, N., López-Pérez, A., Ríos-Jara, E., Ahumada-Sempoal, M. A., Ortiz, M., & Rodríguez-Zaragoza, F. A. (2017). Ecological and functional diversity of gastropods associated with hermatypic corals of the Mexican tropical Pacific. Mar. Biodiv. https://doi.org/10.1007/s12526-017-0780-6

Balbar, A., & Metaxas, A. (2019). The current application of ecological connectivity in the design of marine protected areas. *Global Ecology and Conservation, 17*, e00569.

Bhattacharya, D., Agrawal, S., Aranda, M., Baumgarten, S., Belcaid, M., Drake, J. L., et al. (2016). Comparative genomics explains the evolutionary success of reef-forming corals. *eLife, 5*, e13288.

Breedy, O., & Guzman, H. M. (2011). A revision of the genus *Heterogorgia* Verrill, 1868 (Anthozoa: Octocorallia: Plexauridae). *Zootaxa, 2995*, 27–44.

Breedy, O., & Guzmán, H. M. (2007). A revision of the genus Leptogorgia Milne Edwards & Haime, 1857 (Coelenterata: Octocorallia: Gorgoniidae) in the eastern Pacific. *Zootaxa, 1419*, 1–90.

Breedy, O., Guzman, H. M., & Vargas, S. (2009). A revision of the genus *Eugorgia* Verrill, 1868 (Coelenterata: Octocorallia: Gorgoniidae). *Zootaxa, 2151*, 1–46.

Brill, R. W. (1996). Selective advantages conferred by the high-performance physiology of tunas, billfishes, and dolphin fish. *Comparative Biochemistry and Physiology, 113*, 3–15.

Burke, L., Reytar, K., Spalding, M. & A. Perry, 2011. Reefs at risk revisited. World Resources Institute.

Cabral-Tena, R. A., López-Pérez, A., Reyes-Bonilla, H., Calderon-Aguilera, L. E., Norzagaray-López, C. O., Rodríguez-Zaragoza, F. A., et al. (2018). Calcification of coral assemblages in the eastern Pacific: Reshuffling calcification scenarios under climate change. *Ecological Indicators, 95*, 726–734.

Cabral-Tena, R. A., Reyes-Bonilla, H., Lluch-Cota, S., Paz-García, D., Calderón-Aguilera, L., Norzagaray-López, O., et al. (2013). Different calcification rates in males and females of the coral *Porites panamensis* in the Gulf of California. *Marine Ecology Progress Series, 476*, 1–8.

Cairns, S. D., & Fautin, D. G. (2009). Cnidaria: Introduction. In D. L. Felder & D. K. Camp (Eds.), *Gulf of Mexico origin, waters, and biota* (pp. 315–319). Texas A&M University Press.

Cairns, S. D., Jaap, W. C., & Lang, J. C. (2009). Scleractinia (Cnidaria) of the Gulf of Mexico. In D. L. Felder & D. K. Camp (Eds.), *Gulf of Mexico origin, waters, and biota* (pp. 333–347). Texas A&M University Press.

Calder, D. R., & Cairns, S. D. (2009). Hydroids (Cnidaria: Hydrozoa) of the Gulf of Mexico. In D. L. Felder & D. K. Camp (Eds.), *Gulf of Mexico origin, waters, and biota* (pp. 381–394). Texas A&M University Press.

Calderon-Aguilera, L., Rivera-Monroy, V., Porter-Bolland, L., Martínez-Yrízar, A., Ladah, L., Martínez-Ramos, M., et al. (2012). An assessment of natural and human disturbance effects on Mexican ecosystems: current trends and research gaps. *Biodiversity and Conservation, 21*, 1–29. https://doi.org/10.1007/s10531-011-0218-6.

Calderon-Aguilera, L. E., & Reyes-Bonilla, H. (2016). A comparative network analysis of a marginal reef from the eastern pacific prior and after 20 years of protection. In *Proceedings of the 13th International Coral Reef Symposium*, Honolulu, 19–24 June 2016.

Calderon-Aguilera, L. E., Reyes-Bonilla, H., & Carriquiry-Beltran, J. D. (2007). El papel de los arrecifes coralinos en el flujo de carbono en el océano: estudios en el Pacífico mexicano. In G. Gaxiola-Castro & B. Hernández (Eds.), *Carbono en ecosistemas acuáticos de México* (pp. 215–226). Instituto Nacional de Ecología.

Camarena-Luhrs, T., Gray-Vallejo, E., Liaño-Carrera, F., & Aragón-González, J. (2017). Localización y Superficies de los Principales Edificios Arrecifales y Definición de Grupos Morfo-Funcionales en el Parque Nacional Sistema Arrecifal Veracruzano. Memorias X Congreso de Áreas Protegidas. Julio 3–7. La Habana, Cuba (pp. 339–368).

Cepeda-Morales, J., Beier, E., Gaxiola-Castro, G., Lavín, M. F., & Godínez, V. M. (2009). Effect of the oxygen minimum zone on the second chlorophyll maximum in the Eastern Tropical Pacific off Mexico. *Ciencias Marinas, 35*, 389–403.

Chapa-Balcorta, C., Hernández-Ayón, J. M., Durazo, R., Beier, E., Alin, S. R., & López-Pérez, A. (2015). Influence of post-Tehuano oceanographic processes in the dynamics of the CO2 system in the Gulf of Tehuantepec, Mexico. *Journal of Geophysical Research, 120*(12), 7752–7770.

Chávez-Romo, H. E., Correa-Sandoval, F., Paz-García, D. A., Reyes-Bonilla, H., López-Pérez, R. A., Medina-Rosas, P., et al. (2009). Genetic structure of the scleractinian coral, Pocillopora damicornis, from the Mexican Pacific. In *Proceedings of the 11th International Coral Reef Symposium* (Vol. 1, pp. 429–433).

Cohen, A. L., & Holcomb, M. (2009). Why corals care about ocean acidification. *Oceanography, 22*, 118–127.

Costello, M. J., Coll, M., Danovaro, R., Halpin, P., Ojaveer, H., & Miloslavich, P. (2010). A census of marine biodiversity knowledge, resources, and future challenges. *PLoS One, 5*(8), e12110.

Crawley, A., Kline, D., Dunn, S., Anthony, K., & Dove, S. (2010). The effect of ocean acidification on symbiont photorespiration and productivity in Acropora formosa. *Global Change Biology, 16*, 851–863.

Darnell, R. M., & Defenbaugh, R. E. (1990). Gulf of Mexico: Environmental overview and History of Environmental Research. *American Zoologist, 30*, 3–6.

Duffy, J. E., Amaral-Zettler, L. A., Fautin, D. G., Paulay, G., Rynearson, T. A., Sosik, H. M., et al. (2013). Envisioning a marine biodiversity observation network. *Bioscience, 63*(5), 350–361.

Enochs, I. C., & Glynn, P. (2017). Trophodynamics of Eastern Pacific Coral Reefs. In W. P. Glynn, D. P. Manzello, & I. C. Enochs (Eds.), *Coral Reefs of the Eastern Tropical Pacific* (pp. 291–314). New York, EE. UU.: Springer. https://doi.org/10.1007/978-94-017-7499-4_9.

Fauchald, K., Granados-Barba, A., & Solís-Weiss, V. (2009). Polychaeta (Annelida) of the Gulf of Mexico. In D. L. Felder & D. K. Camp (Eds.), *Gulf of Mexico origin, waters, and biota* (pp. 751–788). Texas A&M University Press.

Fautin, D. G., & Daly, M. (2009). Actinaria, Corallimorpharia, and Zoanthidea (Cnidaria) of the Gulf of Mexico. In D. L. Felder & D. K. Camp (Eds.), *Gulf of Mexico origin, waters, and biota* (pp. 349–357). Texas A&M University Press.

Feely, R. A., Doney, S. C., & Cooley, S. R. (2009). Ocean acidification: Present conditions and future changes in a high-CO2 world. *Oceanography, 22*, 37–47.

Felder, D. L., Álvarez, F., Goy, J. W., & Lemaitre, R. (2009). Decapoda (Crustacea) of the Gulf of Mexico, with comments on the Amphionidacea. In D. L. Felder & D. K. Camp (Eds.), *Gulf of Mexico origin, waters, and biota* (pp. 1019–1104). Texas A&M University Press.

Flores-Rodríguez, P., Flores-Garza, R., García-Ibáñez, S., Torreblanca-Ramírez, C., Galeana-Rebolledo, L., & Santiago-Cortes, E. (2014). Mollusks of the Rocky Intertidal Zone at Three Sites in Oaxaca, Mexico. *Open Journal of Marine Science, 4*, 326–337.

Franco, A. C., Hernandez-Ayon, J. M., Beier, E., Garcon, E., Maske, H., Paulmier, A., et al. (2014). Air-sea CO2 fluxes above the stratified oxygen minimum zone in the coastal region off Mexico. *Journal of Geophysical Research: Oceans, 119*. https://doi.org/10.1002/2013JC009337.

Fredericq, S., Cho, T. O., Earle, S. A., Gurgel, C. F., Krayesky, D. M., Mateo-Cid, L. E., et al. (2009). Seaweeds of the Gulf of Mexico. In D. L. Felder & D. K. Camp (Eds.), *Gulf of Mexico origin, waters, and biota* (pp. 187–259). Texas A&M University Press.

García-Madrigal, M. S., & Andréu-Sánchez, L. I. (2009). Los cangrejos porcelánidos (Decapoda: Anomura) del Pacífico sur de México, incluyendo una lista y clave de identificación para todas las especies del Pacífico oriental tropical. *Ciencia y Mar*, 13, 23–54.

Garcés-Rodríguez, Y., Sánchez-Velasco, L., Díaz-Viloria, N., Jiménez-Rosenberg, S. P. A., Godínez, V., Montes-Arechiga, J., et al. (2018). Larval distribution and connectivity of the endemic Sciaenidae species in the Upper Gulf of California. *Journal of Plankton Research, 40*, 606–618.

Gilly, W., Beman, J. M., Litvin, S. Y., & Robinson, B. H. (2013). Oceanographic and biological effects of shoaling of the Oxygen Minimum Zone. *Annual Review of Marine Science, 5*, 393–420. https://doi.org/10.1146/annurev-marine-120710-100849.

Gittings, S. R. (2009). Cirripedia (Crustacea) of the Gulf of Mexico. In D. L. Felder & D. K. Camp (Eds.), *Gulf of Mexico origin, waters, and biota* (pp. 827–836). Texas A&M University Press.

Glynn, P. W., Alvarado, J. J., Banks, S., Cortés, J., Feingold, J. S., Jiménez, C., et al. (2017). Eastern Pacific coral reef provinces, coral community structure and composition: An overview. In *Coral Reefs of the Eastern Tropical Pacific* (pp. 107–176). Springer.

Glynn, P. W., & Morales, G. E. L. (1997). Coral reefs of Huatulco, West Mexico: reef development in upwelling Gulf of Tehuantepec. *Revista de Biología Tropical*, 1033–1047.

González-Barrios, F. J., & Álvarez-Filip, L. (2018). A framework for measuring coral species-specific contribution to reef functioning in the Caribbean. *Ecological indicators, 95*, 877–886.

Granja-Fernández, R., Herrero-Pérezrul, M. D., López-Pérez, R. A., Hernández, L., Rodríguez-Zaragoza, F. A., Jones, R. W., et al. (2014). Ophiuroidea (Echinodermata) from coral reefs in the Mexican Pacific. *ZooKeys, 406*, 101.

Heard, R. W., & Anderson, G. (2009). Tanaidace (Crustacea) of the Gulf of Mexico. In D. L. Felder & D. K. Camp (Eds.), *Gulf of Mexico origin, waters, and biota* (pp. 987–1000). Texas A&M University Press.

Heard, R. W., & Roccatagliata, D. (2009). Cumacea (Crustacea) of the Gulf of Mexico. In D. L. Felder & D. K. Camp (Eds.), *Gulf of Mexico origin, waters, and biota* (pp. 1001–1011). Texas A&M University Press.

Hernández, L., Balart, E. F., & Reyes-Bonilla, H. (2009). Checklist of reef decapod crustaceans (Crustacea: Decapoda) in the southern Gulf of California, México. Zootaxa, 2119, 39–50.

Hernández, L., Reyes-Bonilla, H., & Balart, E. F. (2010). Effect of coral bleaching induced by low temperature on reef-associated decapod crustaceans of the southwestern Gulf of California. Revista Mexicana de Biodiversidad, 81, S113–S119.

Hernández, L., Ramírez-Ortiz, G., & Reyes-Bonilla, H. (2013). Coral-associated decapods (Crustacea) from the Mexican Tropical Pacific coast. Zootaxa, 3609, 451–464.

Heilprin, A. (1890). The corals and coral reefs of the western waters of the Gulf of Mexico. *Proceedings of the Academy of Natural Sciences of Philadelphia, 42*, 303–316.

Hernández-Ayón, J. M., Zirino-Weiss, A., Delgadillo-Hinojosa, F., & Galindo-Bect, S. (2007). Carbono inorgánico disuelto en el Golfo de California en condiciones de verano. In B. H. de la Torre & G. Gaxiola-Castro (Eds.), *Carbono en ecosistemas acuáticos de México* (pp. 45–57). Instituto Nacional de Ecología, Semarnat and Centro de Investigación Científica y de Educación Superior de Ensenada.

Hoegh-Guldberg, O., Mumby, P. J., Hooten, A. J., Steneck, R. S., Greenfield, P., Gomez, E., et al. (2007). Coral reefs under rapid climate change and ocean acidification. *Science, 318*, 1737–1742.

Hofmann, G. E., Barry, J. P., Edmunds, P. J., Gates, R. D., Hutchins, D. A., Klinger, T., et al. (2010). The effect of ocean acidification on calcifying organisms in marine ecosystems: an organism-to-ecosystem perspective. *Annual Review of Ecology, Evolution, and Systematics, 41*, 127–147.

Horta-Puga, G., Vargas-Hernández, J. M., & Carricart-Ganivet, J. P. (2007). Reef corals. In J. W. Tunnell Jr., E. Chávez, & K. Withers (Eds.), *Coral Reefs of the Southern Gulf of México* (pp. 95–101). Texas A&M University Press.

Hughes, T. P., Barnes, M. L., Bellwood, D. R., Cinner, J. E., Cumming, G. S., Jackson, J. B., et al. (2017). Coral reefs in the Anthropocene. *Nature, 546*(7656), 82–90.

Hughes, T. P., Kerry, J. T., Baird, A. H., Connolly, S. R., Dietzel, A., Eakin, C. M., et al. (2018). Global warming transforms coral reef assemblages. *Nature, 556*(7702), 492.

Humara-Gil, K. J., & Cruz-Gomez, C. (2018). New records of benthic hydroids (Cnidaria: Hydrozoa) from the coast of Oaxaca, Mexico. *Zootaxa, 4455*(3), 454–470.

Hutchins, D. A., Mulholland, M. R., & Fu, F. X. (2009). Nutrient cycles and marine microbes in a CO2-enriched ocean. *Oceanography, 22*, 128–145.

Iglesias-Rodriguez, M. D., Halloran, P. R., Rickaby, R. E. M., Hall, I. R., Colmenero-Hidalgo, E., Gittins, J. R., et al. (2008). Phytoplankton calcification in a high-CO2 world. *Science, 320*, 336–340.

Jackson, J., Donovan, M. A. R. Y., Cramer, K., & Lam, V. (2014). *Status and trends of Caribbean coral reefs: 1970–2012. Global coral reef monitoring network*. Washington, DC: International Union for the Conservation of Nature Global Marine and Polar Program.

Jarquín-González, J., & García-Madrigal, M. D. S. (2010). Tanaidáceos (Crustacea: Peracarida) de los litorales de Guerrero y Oaxaca, México. *Revista Mexicana de Biodiversidad, 81*, 51–61.

Johnston, M. W., & Bernard, A. M. (2017). A bank divided: quantifying a spatial and temporal connectivity break between the Campeche Bank and the northeastern Gulf of Mexico. *Marine Biology, 164*. https://doi.org/10.1007/s00227-016-3038-0.

Jordán-Dahlgren, E. (1993). Atlas de los arrecifes coralinos del Caribe mexicano (No. G 1546. L1. J67 1993).

Jordán-Dahlgren, E. (2002). Gorgonian distribution patterns in coral reef environments of the Gulf of Mexico: Evidence of sporadic ecological connectivity. *Coral Reefs, 21*, 205–215.

Jordán-Dahlgren, E., & Rodríguez-Martínez, R. E. (2003). The Atlantic coral reefs of Mexico. In J. Cortés (Ed.), *Latin America Coral Reefs* (pp. 131–158). Elsevier.

Judkins H L, Vecchione M. & CFE Roper, 2009. Cephalopoda (Mollusca) of the Gulf of Mexico. In: Felder D L and Camp D K (eds). Gulf of Mexico origin, waters, and biota. Texas A&M University Press. 701-709.

Karstensen, J., Stramma, L., & Visbeck, M. (2008). Oxygen minimum zones in the eastern tropical Atlantic and Pacific oceans. *Progress in Oceanography, 77*, 331–350.

Keeling, R. F., Kortzinger, A., & Gruber, N. (2010). Ocean deoxygenation in a warming world. *Annual Review of Marine Science, 2*, 463–493. https://doi.org/10.1146/annurev.marine.010908.163855.

Ketchum J. T., & Reyes-Bonilla H. (2001). Taxonomıa y distribucion de los corales hermat ıpicos (Scleractinia) del Archipielago de Revillagigedo, Pacıfico de Mexico. *Revista de Biologıa Tropical, 49*, 803–848.

Kurihara, H. (2008). Effects ofCO2-driven ocean acidification on the early developmental stages of invertebrates. *Marine Ecology Progress Series, 373*, 275–284.

Lavín, M. F., & Marinone, S. G. (2003). An overview of the physical oceanography of the Gulf of California. In *Nonlinear processes in geophysical fluid dynamics* (pp. 173–204). Dordrecht: Springer.

LeCroy, S. E., Gasca, R., Winfield, I., Ortiz, M., & Escobar-Briones, E. (2009). Amphipoda (Crustacea) of the Gulf of Mexico. In D. L. Felder & D. K. Camp (Eds.), *Gulf of Mexico origin, waters, and biota* (pp. 941–972). Texas A&M University Press.

León-Pech, M. G., Cruz-Barraza, J. A., Carballo, J. L., Calderon-Aguilera, L. E., & Rocha-Olivares, A. (2015). Pervasive genetic structure at different geographic scales in the coral-excavating sponge Cliona vermifera (Hancock, 1867) in the Mexican Pacific. *Coral Reefs, 34*, 887–897.

Lequeux, B. D., Ahumada-Sempoal, M. A., López-Pérez, A., & Reyes-Hernández, C. (2018). Coral connectivity between equatorial eastern Pacific marine protected areas: A biophysical modeling approach. *PLoS One*, e0202995.

López-Pérez, A., Cupul-Magaña, A., Ahumada-Sempoal, M., Medina-Rosas, P., Reyes-Bonilla, H., Herrero-Pérezrul, M. D., et al. (2015). The coral communities of the Islas Marias archipelago, Mexico: Structure and biogeographic relevance to the eastern Pacific. *Marine Ecology, 37*, 679–690.

López-Pérez, A., Granja-Fernández, R., Aparicio-Cid, C., Zepeta-Vilchis, R. C., Torres-Huerta, A. M., Benítez-Villalobos, F., et al. (2014). Corales pétreos, equinodermos y peces asociados a comunidades y arrecifes coralinos del Parque Nacional Huatulco, Pacífico sur Mexicano. *Revista Mexicana de Biodiversidad, 85*(4), 1145–1159.

López-Pérez, A., & López-López, D. A. (2016). Bioerosive impact of Diadema mexicanum on southern Mexican Pacific coral reefs. *Ciencias Marina, 42*(1), 67–79.

López-Pérez, R. A., & Hernández-Ballesteros, L. M. (2004). Coral community structure and dynamics in the Huatulco area, western Mexico. *Bulletin of Marince Science, 75*(3), 453–472.

Manzello, D. P., Enochs, C., Bruckner, A., Renaud, P. G., Kolodziej, G., Budd, D., et al. (2014). Galápagos coral reef persistence after ENSO warming across an acidification gradient. *Geophysical Research Letters, 41*, 9001–9008. https://doi.org/10.1002/2014GL062501.

Manzello, D. P., Enochs, C., Melo, N., Gledhill, D. K., & Johns, E. M. (2012). Ocean acidification refugia of the Florida Reef Tract. *PLoS One, 7*, e41715. https://doi.org/10.1371/journal.pone.0041715.

Manzello, D. P., Kleypas, J. A., Budd, D. A., Eakin, C. M., Glynn, P. W., & Langdon, C. (2008). Poorly cemented coral reefs of the eastern tropical Pacific: Possible insights into reef development in a high-CO2 world. *Proceedings of the National Academy of Sciences*.

Marinone, S. G., Ulloa, M. J., Parés-Sierra, A., Lavin, M. F., & Cudney-Bueno, R. (2008). Connectivity in the northern Gulf of California from particle tracking in a three-dimensional numerical model. *Journal of Marine System, 71*, 149–158.

Martínez-Castillo, V., Reyes-Bonilla, H., & Rocha-Olivares, A. (2018). High genetic diversity and limited genetic connectivity in 2 populations of an endemic and endangered coral species: Porites sverdrupi. *Ciencias Marinas, 44*, 49–58.

Martínez-Guerrero, B. (2007). Nuevos registros de camarones carideos intermareales (Crustacea: Caridea) de la costa de Oaxaca, México. Pp. 47–53. In ME., Hendrickx (Ed.), Contributions to the study of Eastern Pacific Crustaceans 4(2). Instituto de Ciencias del Mar y Limnologia, 195 pp.

McEachran, J. D. (2009). Fishes (Vertebrata: Pisces) of the Gulf of Mexico. In D. L. Felder & D. K. Camp (Eds.), *Gulf of Mexico origin, waters, and biota* (pp. 1223–1316). Texas A&M University Press.

Medellín-Maldonado, F., Cabral-Tena, R. A., López-Pérez, A., Calderón-Aguilera, L. E., Norzagaray-López, C. O., Chapa-Balcorta, C., et al. (2016). Calcificación de las principales especies de corales constructoras de arrecifes en la costa del Pacífico del sur de México. *Ciencias Marinas, 42*(3), 209–225.

Millero, F. J. (2007). The marine inorganic carbon cycle. *Chemical Review, 107*, 308–341.

Miloslavich, P., Díaz, J. M., Klein, E., Alvarado, J. J., Díaz, C., Gobin, J., et al. (2010). Marine biodiversity in the Caribbean: Regional estimates and distribution patterns. *PLoS One, 5*(8), e11916. https://doi.org/10.1371/journal.pone.0011916.

Moberg, F., & Folke, C. (1999). Ecological goods and services of coral reef ecosystems. *Ecological Economics, 29*(2), 215–233.

Montaño-Cortés, C., Marinone, S. G., & Valenzuela, E. (2017). Three-dimensional connectivity in the Gulf of California: An online interactive webpage. *Latin American Journal of Aquatic Research, 45*, 322–328.

Muehllehner, N., Langdon, C., Venti, A., & Kadko, D. (2016). Dynamics of carbonate chemistry, production, and calcification of the Florida Reef Tract (2009–2010): Evidence for seasonal dissolution. *Global Biogeochemical Cycles, 30*, 661–688. https://doi.org/10.1002/2015GB005327.

Munday, P. L., Dixson, D. L., Donelson, J. M., Jones, G. P., Pratchett, M. S., Devitsina, G. V., et al. (2009). Ocean acidification impairs olfactory discrimination and homing ability of a marine fish. *Proceedings of the National Academy of Sciences, 106*, 1848–1852.

Munguia-Vega, Jackson, A., Marinone, S. G., Erisman, B., Moreno-Baez, M., Girón-Nava, A., et al. (2014). Asymmetric connectivity of spawning aggregations of a commercially important marine fish using a multidisciplinary approach. *PeerJ, 2*, e511. https://doi.org/10.7717/peerj.511.

Murphy, S. J., & Hurlburt, H. E. (1999). The connectivity of eddy variability in the Caribbean Sea, the Gulf of Mexico, and the Atlantic Ocean. *Journal of Geophysical Research, 104*, 1431–1453.

Norzagaray-López, C. O., Hernández-Ayón, J. M., Calderon Aguilera, L. E., Reyes-Bonilla, H., Chapa-Balcorta, C., & Ayala-Bocos, A. (2017). Aragonite saturation and pH variation in a fringing reef are strongly influenced by oceanic conditions. *Limnology and Oceanography, 62*(6), 2375–2388.

Ochoa-Chávez, J. M., Del Río-Portilla, M. A., Calderón-Aguilera, L. E., & Rocha-Olivares, A. (2018). Genetic connectivity of the endangered brown sea cucumber Isostichopus fuscus in the northern Gulf of California revealed by novel microsatellite markers. *Revista Mexicana de Biodiversidad, 89*, 563–567.

Opresko, D. M. (2009). Antipatharia (Cnidaria) of the Gulf of Mexico. In D. L. Felder & D. K. Camp (Eds.), *Gulf of Mexico origin, waters, and biota* (pp. 359–363). Texas A&M University Press.

Orr, J. C., Fabry, V. J., Aumont, O., Bopp, L., Doney, S. C., Feely, R. A., et al. (2005). Anthropogenic ocean acidification over the twenty-first century and its impact on calcifying organisms. *Nature, 437*, 681–686.

Paulmier, A., Ruiz-Pino, D., & Garcon, V. (2008). The oxygen minimum zone (OMZ) off Chile as intense source of CO2 and N2O. *Continental Shelf Research, 28*, 2746–2756. https://doi.org/10.1016/j.csr.2008.09.012.

Pawson D L, Vance D J, Messing C G, Solís-Marin F A & C.L. Mah, 2009. Echinodermata of the Gulf of Mexico. In: Felder D L and Camp D K (eds). Gulf of Mexico origin, waters, and biota. Texas A&M University Press. 1177-1204.

Paz-García, D. A., Chávez-Romo, H. E., Correa-Sandoval, F., Reyes-Bonilla, H., López-Pérez, R. A., Medina-Rosas, P., et al. (2012). Genetic connectivity patterns of corals Pocillopora damicornis and Porites panamensis (Anthozoa: Scleractinia) along the west coast of Mexico. *Pacific Science, 66*(1), 43–61.

Paz-García, D. A., Correa-Sandoval, F., Chávez-Romo, H. E., Reyes-Bonilla, H., López-Pérez, R. A., Medina-Rosas, P., et al. (2009). Genetic structure of the massive coral Porites panamensis (Anthozoa: Scleractinia) from the Mexican Pacific. In *Proceedings of the 11th International Coral Reef Symposium* (Vol. 1, pp. 449–453).

Pedroche, F. F., Silva, P. C., Aguilar-Rosas, L. E., Dreckmann, K. M., & Aguilar-Rosas, R. (2005). Catálogo de las algas marinas bentónicas del Pacífico de México. I. Chlorophycota. Mexicali, Baja California: Universidad Autónoma Metropolitana, University of California, Universidad Autónoma de Baja California. 136

Pedroche, F. F., Silva, P. C., Aguilar-Rosas, L. E., Dreckmann, K. M., & Aguilar-Rosas, R. (2008). Catálogo de las algas marinas bentónicas del Pacífico de México. II. Phaeophycota. Mexicali, Baja California: Universidad Autónoma Metropolitana, University of California, Universidad Autónoma de Baja California. 146 p.

Peguero-Icaza, M., Sánchez-Velasco, L., Lavín, M. F., Marinone, S. G., & Beier, E. (2011). Seasonal changes in connectivity routes among larval fish assemblages in a semi-enclosed sea (Gulf of California). *Journal of Plankton Research, 33*, 517–533.

Pineda, J., Hare, J. A., & Sponaugle, A. (2007). Larval transport and dispersal in the coastal ocean and consequences for population connectivity. *Oceanography, 20*(3), 22–39.

Pollock, F. J., Lamb, J. B., Field, S. N., Heron, S. F., Schaffelke, B., Shedrawi, G., et al. (2014). Sediment and turbidity associated with offshore dredging increase coral disease prevalence on nearby reefs. *PLoS One, 9*(7), e102498.

Portner, H. O. (2008). Ecosystem effects of ocean acidification in times of ocean warming: A physiologist's view. *Marine Ecology Progress Series, 373*, 203–217.

Portner, H. O., & Farrell, A. P. (2008). Physiology and climate change. *Science, 322*, 690–692.

Price W W, Heard R W. 2009. Mysida (Crustacea) of the Gulf of Mexico. In: Felder D L and Camp D K (eds). Gulf of Mexico origin, waters, and biota. Texas A&M University Press. 929-939.

Prince, E. D., & Goodyear, C. P. (2006). Hypoxia-based habitat compression of tropical pelagic fishes. *Fisheries Oceanography, 15*, 451–464. https://doi.org/10.1111/j.1365-2419.2005.00393.x.

Reaka, M. L., Camp, D. K., Álvarez, F., Gracia, A. G., Ortiz, M., & Vázquez-Bader, A. R. (2009). Stomatopoda (Crustacea) of the Gulf of Mexico. In D. L. Felder & D. K. Camp (Eds.), *Gulf of Mexico origin, waters, and biota* (pp. 901–921). Texas A&M University Press.

Reyes-Bonilla, H. (2003). Coral reefs of the Pacific coast of México. In J. Cortés (Ed.), *Latin American coral reefs* (pp. 331–349). Amsterdam: Elsevier Science. https://doi.org/10.1016/B978-044451388-5/50015-1.

Reyes-Bonilla, H., Álvarez del Castillo-Cárdenas, P. A., Calderón-Aguilera, L. E., Erosa-Ricárdez, C. E., Fernández-Rivera Melo, F. J., Frausto, T. C., et al. (2014). Servicios ambientales de arrecifes coralinos: el caso del Parque Nacional Cabo Pulmo, B. C. S. In J. I. Urciaga-García (Ed.), *Desarrollo regional en Baja California Sur. Una perspectiva de los servicios ecosistémicos* (pp. 38–63). La Paz, México: UABCS.

Reyes-Bonilla, H., & Calderon-Aguilera, L. E. (1999). Population density, distribution and consumption rates of three corallivores at Cabo Pulmo Reef, Gulf of California, Mexico. *Marine Ecology, 20*, 347–357. https://doi.org/10.1046/j.1439-0485.1999.2034080.x.

Reyes-Bonilla, H., Calderon-Aguilera, L. E., Cruz-Piñon, G., Medina-Rosas, P., López-Pérez, R. A., Herrero-Pérezrul, M. D., et al. (2005). *Atlas de los corales pétreos (Anthozoa: scleractinia) del Pacífico Mexicano*. Centro de Investigación Científica y de Educación Superior de Ensenada, Comisión Nacional para el Conocimiento y Uso de la Biodiversidad, Consejo Nacional de Ciencia y Tecnología, Universidad de Guadalajara / Centro Universitario de la Costa, Universidad del Mar. 128 pp. ISBN: 970-27-0779.

Reyes-Bonilla, H., Calderón-Aguilera, L. E., Mozqueda-Torres, M. C., & Carriquiry-Beltrán, J. D. (2014). Presupuesto de carbono en arrecifes coralinos de México. *Interciencia, 39*(9).

Reyes-Bonilla, H., & López-Pérez, R. A. (2009). Corals and coral reef communities in the Gulf of California. In A. Johnson & J. Ledezma-Vázquez (Eds.), *Atlas of coastal ecosystems in the western Gulf of California* (pp. 45–57). Tucson: The University of Arizona Press.

Reyes-Bonilla, H., Sinsel-Duarte, F., & Arizpe-Covarrubias, O. (1997). Gorgonias y corales pétreos (Anthozoa: Gorgonacea y Scleractinia) de Cabo Pulmo, México. *Revista de Biología Tropical, 45*, 1439–1443.

Reyes-Gómez A. (2016). The Polyplacophora from the Mexican Pacific. *Festivus*, 1–50.

Riegl, B., Halfar, J., Purkis, S. J., & Godinez-Orta, L. (2007). Sedimentary facies of the eastern Pacific's northernmost reef-like setting (Cabo Pulmo, Mexico). *Marine Geology, 236*, 61–77. https://doi.org/10.1016/j.margeo.2006.09.021.

Rioja-Nieto, R., & Álvarez-Filip, L. (2019). Coral reef systems of the Mexican Caribbean: Status, recent trends and conservation. *Marine Pollution Bulletin, 140*, 616–625.

Rioja-Nieto, R., Garza-Pérez, R., Álvarez-Filip, L., Ismael, M. T., & Cecilia, E. (2019). The Mexican Caribbean: from Xcalak to holbox. In *World Seas: An environmental evaluation* (pp. 637–653). Academic Press.

Ríos-Jara, E., López-Uriarte, E., & Galván-Villa, CM. (2008). Bivalve molluscs from the continental shelf of Jalisco and Colima, Mexican Central Pacific. Amer Malac Bull. *26*, 119–131.

Rixen, T., Jiménez, C., & Cortés, J. (2012). Impact of upwelling events on the sea water carbonate chemistry and dissolved oxygen concentration in the Gulf of Papagayo (Culebra Bay), Costa Rica: Implications for coral reefs. *International Journal of Tropical Biology and Conservation, 60*, 187–195.

Roberts, C. M., McClean, C. J., Veron, J. E. N., Hawkins, J. P., Allen, G. R., McAllister, D. E., et al. (2002). Marine biodiversity hotspots and conservation priorities for tropical reefs. *Science, 295*, 1280–1284.

Robertson, R., & Cramer, K. L. (2014). Defining and dividing the Greater Caribbean: Insights from the biogeography of shorefishes. *PLoS One, 9*(7), e102918.

Robison, B. H. (2009). Conservation of deep pelagic biodiversity. *Conservation Biology, 23*, 847–858.

Rodríguez-Zaragoza, F. A., & Arias-Gonzalez, J. E. (2008). Additive diversity partitioning of reef fishes across multiple spatial scales. *Caribbean Journal of Science, 44*(1), 90–101.

Rodríguez-Zaragoza, F. A., & Arias-Gonzalez, J. E. (2015). Coral biodiversity and bio-construction in the northern sector of the mesoamerican reef system. *Frontiers in Marine Science, 2*, 13.

Romero-Torres, M., Treml, E. A., Acosta, A., & Paz-García, D. A. (2018). The eastern tropical Pacific coral population connectivity and the role of the eastern Pacific barrier. *Scientific Reports*. https://doi.org/10.1038/s41598-018-27644-2.

Rosa, R., & Seibel, B. A. (2008). Synergistic effects of climate-related variables suggest future physiological impairment in an oceanic predator. *Proceedings of the National Academy of Science, 105*, 20776–20780.

Rosenberg, G., Moretzsohn, F., & García, E. F. (2009). Gastropoda (Mollusca) of the Gulf of Mexico. In D. L. Felder & D. K. Camp (Eds.), *Gulf of Mexico origin, waters, and biota* (pp. 579–699). Texas A&M University Press.

Rützler, K., van Soest, R. W. M., & Piantoni, C. (2009). Sponges (Porifera) of the Gulf of Mexico. In D. L. Felder & D. K. Camp (Eds.), *Gulf of Mexico origin, waters, and biota* (pp. 285–313). Texas A&M University Press.

Saavedra-Sotelo, N., Calderón-Aguilera, L., Reyes-Bonilla, H., López-Pérez, R. A., Medina-Rosas, P., & Rocha-Olivares, A. (2011). Limited genetic connectivity of Pavona gigantea in the Mexican Pacific. *Coral Reefs, 30*, 677–686.

Saavedra-Sotelo, N. C., Calderon-Aguilera, L. E., Reyes-Bonilla, H., Paz-García, D. A., López-Pérez, R. A., Cupul-Magaña, A., et al. (2013). Testing the genetic predictions of a biogeographical model in a dominant endemic Eastern Pacific coral (Porites panamensis) using a genetic seascape approach. *Ecology and Evolution, 3*(12), 4070–4091.

Sabine, C. L., Feely, R. A., Gruber, N., Key, R. M., Lee, K., Bullister, J. L., et al. (2004). The oceanic sink for anthropogenic CO2. *Science, 305*, 367–371.

Salas-Monreal, D., Marín-Hernández, M., Salas-Pérez, J. J., Salas-De León, D. A., Monreal-Gómez, M. A., & Pérez-España, H. (2018). Coral reef connectivity within the Western Gulf of Mexico. *Journal of Marine Systems, 197*, 88–99.

Sanvicente-Añorve, L., Zavala-Hidalgo, J., Allende-Arandía, M. E., & Hermoso-Salazar, M. (2014). Connectivity patterns among coral reef systems in the southern Gulf of Mexico. *Marine Ecology Progress Series, 498*, 27–41.

Schmitter-Soto, J. J., Aguilar-Perera, A., Cruz-Martínez, A., Herrera-Pavón, R. L., Morales-Aranda, A. A., & Sobián-Rojas, D. (2018). Interdecadal trends in composition, density, size, and mean trophic level of fish species and guilds before and after coastal development in the Mexican Caribbean. *Biodiversity and Conservation, 27*(2), 459–474.

Schotte, M., Markham, J. C., & Wilson, G. D. F. (2009). Isopoda (Crustacea) of the Gulf of Mexico. In D. L. Felder & D. K. Camp (Eds.), *Gulf of Mexico origin, waters, and biota* (pp. 973–986). Texas A&M University Press.

Shaw, E., McNeil, B., & Tilbrook, B. (2012). Impacts of ocean acidification in naturally variable coral reef flat ecosystems. *Journal of Geophysical Research, 117*, C03038. https://doi.org/10.1029/2011JC007655.

Soria, G., Torre-Cosio, J., Munguia-Vega, A., Marinone, S. G., Lavín, M. F., Cinti, A., et al. (2014). Dynamic connectivity patterns from an insular marine protected área in the Gulf of California. *Journal of Marine Systems, 129*, 248–258.

Spalding, M. D., Fox, H. E., Allen, G. R., Davidson, N., Ferdaña, Z. A., Finlayson, M., et al. (2007). Marine ecoregions of the world: A bioregionalization of coastal and shelf areas. *BioScience, 57*(7), 573–583.

Stramma, L., Johnson, G. C., Sprintall, J., & Mohrholz, V. (2008). Expanding oxygen-minimum zones in the tropical oceans. *Science, 320*, 655–658.

Suchley, A., & Alvarez-Filip, L. (2018). Local human activities limit marine protection efficacy on Caribbean coral reefs. *Conservation Letters, 11*(5), e12571.

Suchley, A., McField, M. D., & Alvarez-Filip, L. (2016). Rapidly increasing macroalgal cover not related to herbivorous fishes on Mesoamerican reefs. *PeerJ, 4*, e2084.

Thomson, D. A., Findley, L. T., & Kerstitch, A. N. (2000). Reef fishes of the Sea of Cortez: The rocky-shore fishes of the Gulf of California. University of Texas Press. 353 p.

Trasviña, A., Barton, E. D., Brown, J., Velez, H. S., Kosro, P. M., & Smith, R. L. (1995). Offshore wind forcing in the Gulf of Tehuantepec, Mexico: The asymmetric circulation. *Journal of Geophysical Research, 100*(C10), 20,649–20,663.

Tunnell, J. W., Jr. (2007). Reefs distribution. In J. W. Tunnell Jr., E. A. Chávez, & K. Withers (Eds.), *Coral reefs of the southern Gulf of Mexico* (pp. 17–29). USA: Texas A&M University Corpus Chisti.

Turgeon, D. D., Lyons, W. G., Mikkelsen, P., Rosenberg, G., & Moretzsohn, F. (2009). Bivalvia (Mollusca) of the Gulf of Mexico. In D. L. Felder & D. K. Camp (Eds.), *Gulf of Mexico origin, waters, and biota* (pp. 711–744). Texas A&M University Press.

Ulloa, O., Canfield, D. E., DeLong, E. F., Letelier, R. M., & Stewart, F. J. (2012). Microbial oceanography of anoxic marine zones. *Proceedings of the National Academy of Science, 109*, 15996–16003. https://doi.org/10.1073/pnas.1205009109.

Valencia-Mendez, O., Lopez-Perez, A., Martinez-Guerrero, B., Antonio-Perez, V., & Ramirez-Chavez, E. (2017). A new record of Harlequin Shrimp (Malacostraca: Decapoda: Palaemonidae: Hymenocera picta Dana, 1852) in the southern Mexican Pacific Reefs. *Journal of Threatened Taxa, 9*(8), 10571–10576.

Vaquer-Sunyer, R., & Duarte, C. M. (2008). Thresholds of hypoxia for marine biodiversity. *Proceedings of the National Academy of Science, 105*, 15452–15457.

Vega Thurber, R. L., Burkepile, D. E., Fuchs, C., Shantz, A. A., McMinds, R., & Zaneveld, J. R. (2014). Chronic nutrient enrichment increases prevalence and severity of coral disease and bleaching. *Global Change Biology, 20*(2), 544–554.

Veron, J. E. N., Hoegh-Guldberg, O., Lenton, T. M., Lough, J. M., Obura, D. O., Pearce-Kelly, P., et al. (2009). The coral reef crisis: the critical importance of <350 ppm CO2. *Marine Pollution Bulletin, 58*, 1428–1436.

Villegas-Sánchez, C. A., Pérez-España, H., Rivera-Madrid, R., Salas-Monreal, D., & Arias-González, J. E. (2014). Subtle genetic connectivity between Mexican Caribbean and south-western Gulf of Mexico reefs: the case of the bicolor damselfish, Stegastes partitus. *Coral Reefs, 33*, 241–251.

Wenger, A. S., Williamson, D. H., da Silva, E. T., Ceccarelli, D. M., Browne, N. K., Petus, C., et al. (2016). Effects of reduced water quality on coral reefs in and out of no-take marine reserves. *Conservation Biology, 30*(1), 142–153.

Wiedenmann, J., D'Angelo, C., Smith, E. G., Hunt, A. N., Legiret, F.-E., Postle, A. D., et al. (2013). Nutrient enrichment can increase the susceptibility of reef corals to bleaching. *Nature Climate Change, 3*, 160–164.

Wood, S., Paris, C. B., Ridgwell, A., & Hendy, E. J. (2014). Modelling dispersal and connectivity of broadcast spawning corals at the global scale. *Global Ecology and Biogeography, 23*, 1–11.

Worm, B., Barbier, E. B., Beaumont, N., Duffy, J. E., Folke, C., Halpern, B. S., et al. (2006). Impacts of biodiversity loss on ocean ecosystem services. *Science, 314*, 787–790.

The Humboldt Current Large Marine Ecosystem (HCLME), a Challenging Scenario for Modelers and Their Contribution for the Manager

Adrien Chevallier, Wolfgang Stotz, Marcel Ramos, and Jaime Mendo

1 Introduction

The ocean provides the major portion of the world's fish products, with ocean-based production accounting for nearly 90% of global landings of capture fish and about a third of aquaculture production (FAO, 2018). As world population and people's incomes rise, the demand for ocean-derived food continues to grow. By some estimates, nearly 500 million tons of animal meat will be required to feed the world's population in 2050 (FAO, 2018), to which the ocean can make a large contribution. At the same time, hunger and malnutrition continues to be a challenge in many countries, especially in rural or developing areas (FAO, 2018). Costello et al. (2019)

A. Chevallier
Grupo de Ecología y Manejo de Recursos (Ecolmar), Departamento de Biología Marina, Facultad de Ciencias del Mar, Universidad Católica del Norte, Coquimbo, Chile

W. Stotz (✉)
Grupo de Ecología y Manejo de Recursos (Ecolmar), Departamento de Biología Marina, Facultad de Ciencias del Mar, Universidad Católica del Norte, Coquimbo, Chile

Centro de Estudios Avanzados en Zonas Áridas (CEAZA), La Serena, Chile
e-mail: wstotz@ucn.cl

M. Ramos
Departamento de Biología Marina, Facultad de Ciencias del Mar, Universidad Católica del Norte, Coquimbo, Chile

Núcleo Milenio de Ecología y Manejo Sustentable de Islas Oceánicas (ESMOI), Universidad Católica del Norte, Coquimbo, Chile

Centro de Estudios Avanzados en Zonas Áridas (CEAZA), Coquimbo, Chile

Centro de Innovación Acuícola Aquapacífico, Coquimbo, Chile

J. Mendo
Departamento Académico de Manejo Pesquero y Medio Ambiente, Facultad de Pesquería, Universidad Nacional Agraria La Molina, Lima, Peru

M. Ortiz, F. Jordán (eds.), *Marine Coastal Ecosystems Modelling and Conservation*, https://doi.org/10.1007/978-3-030-58211-1_2

argue that the ocean can play a unique role in contributing to sustainable food security because of the relatively low gas emissions associated with its production (Hoegh-Guldberg et al., 2019), the particular efficiency of marine animals to process animal feed into human food, which provides essential vitamins, minerals, long-chain omega-3 fatty acids, and other nutrients not found in plant-source foods or other animal proteins (Allison, Delaporte, & Hellebrandt de Silva, 2013; Golden et al., 2016; Kawarazuka & Béné, 2010), through being readily available to most coastal populations, especially in the most productive upwelling areas. Moreover, the area suitable for cultivating food from the sea is not limited by the scarcity of land and water resources, and trade plays an important role in moving ocean food products around the world.

Off Peru and Chile, the Humboldt Current Large Marine Ecosystem (HCLME) covers 95% of the southeast Pacific coast of South America, from around 3 to 58°S of latitude and over 200 nautical miles offshore (Gutierrez, Akester, & Naranjo, 2016). About 65% of the HCLME corresponds to the area of influence of the Humboldt Current, a major Eastern Boundary Upwelling Ecosystem (EBUE) called Humboldt Current System (HCS), which extends from south-central Chile (~42°S) up to northern Peru (~4°S). This area is characterized mainly by a surface circulation towards the equator, a cool nutrient-rich subsurface current towards the pole, which is also very poor in dissolved oxygen, and frequent events of coastal upwelling along the coast (Thiel et al., 2007). The very high fish production in the HCLME has been primarily associated with recurrent upwelling pulses and the predominance of short, thermodynamically efficient food chains (Ryther, 1969). The HCLME is a global center for food security, marine biodiversity, global fishmeal production, and climate regulation (FAO, 2018; Gutierrez et al., 2016; Serra, Akester, Bouchón, & Gutierrez, 2012). However, the frequency and intensity of upwelling within the HCS register strong interannual and seasonal variations along the coast, and so do the fisheries landings (Yáñez et al., 2017a). Periodically, the upwelling that drives the system's productivity is disrupted by Equatorial remote forcing such as strong El Niño–Southern Oscillation (ENSO) events. Consequently, fish abundance and distribution are significantly affected, often leading to stock crashes and cascading social and economic impacts (Arntz & Fahrbach, 1996). Future changes in the structure and functioning and therefore the productivity of the HCLME are expected due to climate variability and global warming, which has already proven to affect various well-studied marine ecosystems (Barange et al., 2018; Masson-Delmotte et al., 2018; Pörtner et al., 2019). Global warming is likely to alter the exchanges of energy and matter between the atmosphere, the ocean, and the continents, modifying pressure gradients and coastal wind fields along with marine currents, sea-surface temperature (SST), and thermal stratification, in addition to spatio-temporal distribution of the coastal outcrops. Nevertheless, it is still unclear how these multiple stressors will impact the productivity and biodiversity of the HCLME. While there is an ongoing debate about the response to global warming (Bertrand, Vögler, & Defeo, 2018; Gutierrez et al., 2016; Zavala et al., 2019), likely effects on the phenology, spatial distributions, and species compositions of primary and secondary producers should be observed. In this context, sustainable management of resource use and ecosystems will increase resilience and enable a better adaptive response to climate variability and climate change (Barange et al., 2018; Mendo et al., in press). In this chapter, time-space heterogeneity of upwelling processes and related determinant oceanographic features of the HCLME will be reviewed and summarized according to the state of the art. The resulting heterogeneity in fisheries landings will then be presented for representative species of pelagic, demersal, and coastal benthic ecosystems. Finally, the interplay of this variability and societal processes will be illustrated by a case study.

1.1 Upwelling and Related Determinant Oceanographic Features

The currents along the west coast of South America move towards the pole and the equator (Fig. 1a), carrying overlapping water masses and weakly mixed layers along the Humboldt Current System (Silva, Rojas, & Fedele, 2009). The eastward flowing West Wind Drift reaches the South American continent around 42–48°S and splits

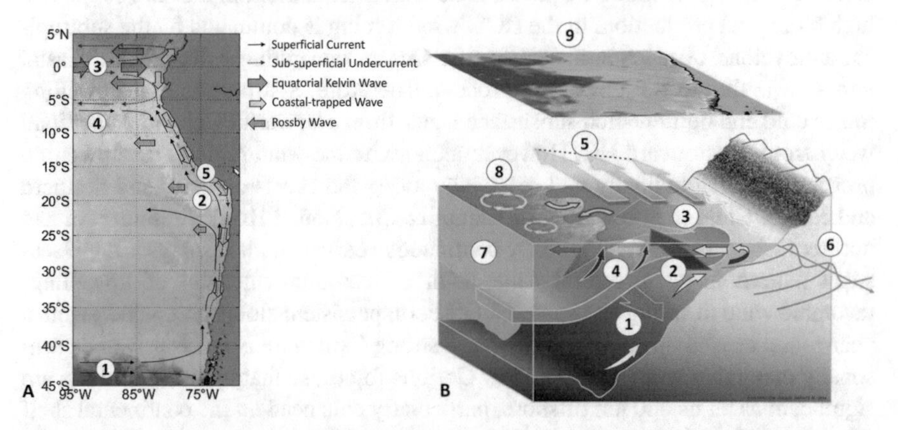

Fig. 1 Biophysical features of the Humboldt Current System (HCS). (**a**) Satellite SST in the Eastern South Pacific during summer of 2015/16 El Niño along with the main currents of the HCS and large-scale wave propagation. 1 West Wind Drift, 2 Humboldt Current, 3 Equatorial Undercurrent, 4 South Equatorial Countercurrent, 5 Peru Chile Undercurrent. (**b**) Upwelling process modulated by Equatorial remote forcing and its effect on primary production based on satellite chlorophyll-a. 1 Peru Chile Undercurrent with nutrient-rich and oxygen depleted sub-superficial waters associated with the Oxygen Minimum Zone, 2 Equatorward Coastal Jet* of the Humboldt Current System, with oxygenated superficial waters, 3 Ekman Transport, 4 Ekman Suction, 5 Wind Stress Curl, 6 Coastal-trapped Waves, 7 Upper mixed layer, 8 Mesoscale Eddy, 9 Mesoscale variability of satellite chlorophyll-a (25–45°S and 70–85°W).* At the surface, the boundary between the upwelled and oceanic waters is often a front, and the upwelling-induced horizontal density gradients support an equatorward coastal jet (Mooers, Collins, & Smith, 1976). Adapted from Gutiérrez et al. (2014), illustrated by Ismael Zarate

into a northward component, the Humboldt Current (HC), and a southward component, the Cape Horn Current (CHC). While the HC transports surface Subantarctic Water towards the equator, the Peru Chile Countercurrent (PCC) transports southward surface Subtropical Water. Subtropical Water is warmer, more saline, and less oxygenated than northward Subantarctic Water. These water masses include a mixed layer around 25–30°S. Below approximatively 100–400 m, the coastal Peru Chile Undercurrent (PCU) carries southward Equatorial Subsurface Water as far as 48°S. This Equatorial Subsurface Water is characterized by a salinity maximum, high nutrient, and low oxygen contents and, by its nutrient-rich composition, plays a major role within the ecosystems of the HCS. The dynamics of these main currents of the HCS is mostly controlled by the atmospheric South Pacific Anticyclone, which is a semi-permanent high-pressure phenomenon in the area.

Underwater, a horizontal layer named the pycnocline separates surface waters, approximately upper 100 meters, from the deep ocean by their large density difference, and hinders vertical transport. However, driven by strong equatorward winds parallel to the coast and strengthening away offshore under a positive gradient, the combination of Ekman suction/pumping and offshore transport produces intense coastal upwelling of subsurface water along the coasts of the Eastern Boundary Upwelling Systems (EBUS) (Fig. 1b) (e.g. Bravo, Ramos, Astudillo, Dewitte, & Goubanova, 2016; Smith, 1968). The vertical mixing that occurs through wind-driven upwelling fertilizes the photic zone with high nutrient inputs and supports a high biological production. In the HCS, wind forcing is dominated by the subtropical anticyclone of the South East Pacific Ocean, generating equatorward coastal winds favorable to the upwelling process. Therefore, a strong coastal upwelling mixes cold and nutrient-rich subsurface water from poleward PCU with superficial water from equatorward HC. However, along with the wind regime, the upwelling process varies latitudinally and seasonally along the coasts of Peru, and northern and central Chile (e.g. reviewed in Kämpf & Chapman, 2016). While in Peru and northern Chile, upwelling is mostly continuous year-round, it displays a more seasonal pattern in south-central Chile with a maximum intensity of upwelling-favorable wind in spring and summer. Off Peru, persistent alongshore wind favors a year-round coastal upwelling. In winter, a strong Ekman transport is driven by seasonally maximum wind stress and low Coriolis forces, so that upwelling effects are significant as far as 400 km offshore, particularly enhanced on the continental shelf near Pisco (13.7°S). Along the Chilean coast, upwelling happens around local spots which are separated by long stretches of coast with or without intermittent and less intense upwelling. Moreover, most prominent coastline features highly enhanced the wind-induced upwelling in four regions of the country, as demonstrated by Figueroa and Moffat (2000): Antofagasta and Mejillones Peninsula (23°S), Punta Lengua de Vaca around Coquimbo Bay (30°S), Valparaíso (33°S), and Punta Lavapié around the Bay of Concepción (37°S).

The nutrient enrichment of the photic zone through upwelling supports a high primary production within the EBUS. This huge sea-surface phytoplankton biomass fuels abundant populations of small pelagic grazers and other trophic groups from the pelagic, demersal, and benthic interconnected food webs, which ultimately sustain coastal fisheries. The northern HCS off Peru is notably more productive than any region in the world in terms of fish per unit area (Chavez, Bertrand, Guevara-Carrasco, Soler, & Csirke, 2008; Fréon, Barange, & Aristegui, 2009). Nonetheless, the regulation of the primary production in each one of the main EBUS is still poorly understood (Messié & Chavez, 2015). Key achievements have been made during the last two decades, as the spatiotemporal dynamics of the primary production has been estimated by satellite imagery (e.g. Carr & Kearns, 2003; Demarcq, 2009; Lara et al., 2019). All along the coast of the HCS, the highest primary production occurs during the spring-summer period, with the highest light intensity. Along the Peruvian coast and south-central Chile (34–37°S), primary production is much higher than in northern Chile mainly due to the presence of a wider continental shelf associated with high intensity of upwelling-favorable winds and great availability of limiting nutrients such as nitrate and iron. Moreover, a significant seasonal trend is observed in the southern HCS with a high contrasting production in spring and summer, mostly due to the maximum intensity of upwelling-favorable winds. In winter, primary production is more influenced by the additional nutrient supply provided by turbid river plumes of major Andean rivers (e.g. Iriarte, Vargas, Tapia, Bermúdez, & Urrutia, 2012). At the local scale, peaks of primary production are associated with main upwelling spots along the Chilean coast: Peninsula Mejillones (~23°S), Punta Lengua de Vaca (30°S), and Punta Lavapié (37°S). A peak was also recently located in the upwelling center of Punta Galera (40°S) (Lara et al., 2019; Pinochet, Garcés-Vargas, Lara, & Olguín, 2019).

In the HCS, the high biological production at the surface leads to heavy organic matter sinking. In the water column, high-rate microbial decomposition of that organic matter consumes the dissolved oxygen. Along with stratification and weak subsurface circulation, it produces the subsurface oxygen minimum zone (OMZ; e.g. Wyrtki, 1962; Helly & Levin, 2004). Dissolved oxygen (DO) in the OMZ reaches values lower than 20 $\mu mol\ l^{-1}$ in the suboxia range, and even anoxic zones may be found (e.g. Ulloa, Canfield, DeLong, Letelier, & Stewart, 2012). These extremely low concentrations largely impact marine life and biogeochemical cycling processes. Through high-resolution simulations, regional patterns of the OMZ dynamics can be modeled despite the lack of observational data (e.g. Montes et al., 2014; Pizarro-Koch et al., 2019; Vergara et al., 2016). According to these models, the oxygen depletion is the most intense from about 5 to 15°S, near the Peruvian coast, with a maximum vertical thickness of about 500 m between 100 and 600 m depth in the suboxia range (DO $<20\ \mu mol\ l^{-1}$, Montes et al., 2014). In this region, the OMZ presents a tongue-like shape extending offshore and seems largely influenced by the equatorial current system. Off Peru, the OMZ appears to have a strong seasonal pattern with minimum DO reached in summer due to a high production of organic matter, and a yearly peak in August during the austral winter associated with increased mixing peaking in July and an intensification of the Peru Chile Undercurrent (PCU) starting in June (Vergara et al., 2016). In contrast to the Peruvian HCS, the influence of the equatorial current system is subdued along the Chilean coast where the subsurface circulation is dominated by the PCU. Offshore extension and thickness of the OMZ should rapidly decrease southward because of the mixing with well-ventilated water masses rising from Antarctica (Pizarro-Koch et al., 2019). However, near the coast, it should expand at intermediate depth as far south as 38°S, during spring-summer with the intensification of the PCU and upwelling summer peak and retract in winter along with PCU intensity.

Each one of the interconnected features of the HCS mentioned above (current system, upwelling events, primary production, and oxygen minimum zone) is regulated by large-scale climatic forcing on a variety of superimposed time scales (summarized in Table 1 and explained in detail hereafter).

More than any other EBUS, the HCS is highly influenced by the propagation of eastward equatorial Kelvin waves (EKW), from intra-seasonal to interdecadal scales. This remote ocean forcing is associated with positive eastward wind stress anomalies in the Central Pacific. On the intra-seasonal scale (~60–120 days), the

Table 1 Large-scale superimposed climatic forcing

Time scale	Forcing	Major known effects on the HCS
Intra-seasonal	– Equatorial and coastal-trapped Kelvin waves (EKW—CTW) – Wind stress variability – Mesoscale structures (eddies, filaments, and fronts)	– Vertical displacements of the pycnocline and thermocline – Modulation of vertical mixing and primary production – Modulation of the offshore advection of phytoplankton-rich coastal waters – Presence of shears of the coastal currents, thermal gradients, and sea level anomalies
Interannual	El Niño–Southern Oscillation (ENSO)	– Presence of temperature and salinity anomalies in the upper ocean layer during ENSO events – Modulation of the amplitude of the Equatorial and coastal-trapped Kelvin waves (EKW–CTW) – Pycnocline, thermocline, and OMZ deepen/shoal during El Niño/ La Niña events – Modulation of the timing and duration of the upwelling seasons in the southern HCS
Interdecadal	Pacific Decadal Oscillation (PDO)	– Increase/decrease in sea-surface temperature anomalies – Pycnocline, thermocline, and sea level slopes higher/lower during warm/cold regimes – Modulation of the vertical mixing and primary production – Ecosystem regime shifts
Long-term	Climate change	– Increase in ocean stratification, deoxygenation, and acidification – Alteration of the alongshore wind stress, upwelling process, and primary production – Projected increase in strong EP El Niño and strong La Niña events – Average sea level rise – Increased intensity and frequency of extreme events – Strong impacts on marine ecosystems

eastward propagation of downwelling equatorial Kelvin waves (EKW), forced by westerly wind, bursts in the western tropical Pacific. When reaching the coasts of Ecuador and Peru, these Kelvin waves generate coastal-trapped waves (CTW) which propagate poleward along the HCS, even as south as 37°S off Chile (e.g. Echevin et al., 2014; Hormazabal, Shaffer, & Pizarro, 2002; Shaffer, Pizarro, Djurfeldt, Salinas, & Rutllant, 1997; Sobarzo et al., 2016). During their propagation, CTW produce vertical displacements of the nearshore pycnocline over tens of meters, which modulate the enrichment of the euphotic layer during upwelling (Fig. 1b) and impact the biological productivity of the coastal system. Furthermore, CTW significantly amplify the core flow of the PCU and have an influence on the westward Rossby waves (Clarke & Shi, 1991; Pizarro, Shaffer, Dewitte, & Ramos, 2002), which play a crucial role with associated currents and eddies in transporting the phytoplankton-rich coastal waters offshore (Bonhomme, Aumont, & Echevin, 2007). Aside from the remote forcing of CTW, events of surface wind intensification off Peru may induce local upwelling on a shorter intra-seasonal scale (~10–60 days). They partly depend on the South Pacific Anticyclone meridional displacements. These atmospheric events also modulate the CTW impacts on the coastal system. In the southern HCS (from about 30°S to 42°S), meridional wind stress strongly enhances upwelling events and phytoplankton production on the intra-seasonal scale (~30–90 days) with a seasonal peak in spring-summer concurrent with maximum solar radiation (e.g. Gomez, Spitz, Batchelder, & Correa-Ramirez, 2017). The wind-driven variability of chlorophyll concentration in the southern HCS is connected to mid- and high latitude atmospheric anomalies which seem to be linked to the tropical Madden-Julian Oscillation (Gomez et al., 2017), but so far, little information is available on the subject.

The coastal intra-seasonal variability is associated with strong spatial heterogeneity on mesoscale and submesoscale, such as eddies, filaments, and fronts (e.g. reviewed by McGillicuddy, 2016). These patterns primarily originate from baroclinic instabilities of the mean flow of surface and subsurface currents, i.e. by misalignment of the pressure and density gradients within the corresponding water masses. They are associated with strong horizontal and vertical shear of the coastal currents, thermal gradients, and sea level anomalies. Mesoscale structures modulate upwelling events, phytoplankton distribution (Fig. 1b), and community structure by processes of horizontal advection, vertical pumping, and trapping of nutrient and phytoplankton, but also by their influence on stratification and therefore on upper ocean mixing. At the interface between coastal and offshore ecosystems, mesoscale motions play a critical role by stirring upwelling-derived biomass offshore and reducing biomass in the upwelling zone. On the coastal margin, they modulate the population dynamics of numerous species by processes of retention and dispersion of phytoplankton and zooplankton (McGillicuddy, 2016), including seaweed propagules and pelagic larvae of upper trophic levels.

On an interannual scale, the El Niño–Southern Oscillation phenomenon (ENSO) overlaps with these regional patterns. The ENSO is a coupled mode of variability of the tropical ocean-atmosphere system. Centered in the Pacific Ocean, ENSO events have global climatic teleconnections and constitute the world most dominant phenomenon on sub-seasonal to interannual time scale (e.g. Yeh et al., 2009). During the warm (El Niño) and cold (La Niña) phases of ENSO events, the physical and biogeochemical features of the northern HCS, as well as the large-scale circulation and ecosystems, have been shown to be strongly impacted (e.g. reviewed by Wang, Deser, Yu, DiNezio, & Clement, 2017). On average, from 1960 to 2014 off the coasts of Peru, large temperature (± 3–4 °C) and salinity (± 0.1–0.2) anomalies were observed down to 100–200 m depth during ENSO events (e.g. 2015–2016 El Niño Fig. 1a) (Grados, Chaigneau, Echevin, & Dominguez, 2018). The thermocline—the thermal layer in which the temperature gradient decreases rapidly with depth and which separates the upper mixed layer from the deep-water masses—deepens of 60 m at 100 km from the coast during El Niño and shoals of 25 m during La Niña events. Indeed, ENSO events modulate the amplitude of intra-seasonal EKW and consequently of CTW. During strong arising El Niño events, the amplitude of the CTW is increased so that the nearshore pycnocline and thermocline deepen by several tens of meters (e.g. Echevin et al., 2014). Vertical mixing is attenuated leading to a reduction in nutrient exchange. Phytoplankton in the upper layer relies on vertical nutrient transport to sustain its productivity, so that the intensified stratification during El Niño events is accompanied by a decreasing primary production (Behrenfeld et al., 2006). Moreover, oxygenation is enhanced in the upper zone as the OMZ deepens through the propagation of downwelling Kelvin waves (e.g. Graco et al., 2017). Conversely, surface cooling during strong arising La Niña events favors elevated vertical exchange and is associated with an increase in primary production, whereas the OMZ rises by tens of meters. However, at least two types of ENSO events occur in the tropical Pacific: the well-documented Eastern-Pacific type that has maximum SST anomalies centered over the eastern tropical Pacific region, and the Central-Pacific (CP) type that has the anomalies centered near the International Date Line (0°S–180°E) (e.g. Yu, Kao, Lee, & Kim, 2011) and therefore impacts the northern HCS with distinctive patterns. Indeed, during CP El Niño, the amplitude of the downwelling EKW reaching the South American coast is much weaker and slightly cooler conditions are observed associated with shallow thermocline (Dewitte et al., 2012). There is a tendency for CP El Niño to occur more often over the last decades. Moreover, on the southern tip of the HCS, the ENSO phenomenon forces biophysical parameters through atmospheric teleconnections (e.g. Montecinos & Gomez, 2010). In particular, the ENSO events modulate the timing and duration of the upwelling seasons in the southern HCS.

In addition to the interannual variability associated with the ENSO phenomenon, multidecadal fluctuations of the main features of the equatorial and tropical Pacific, such as sea temperature anomalies, pycnocline, thermocline, and sea level slopes, are linked to the Pacific Decadal Oscillation (PDO) which occurs on the Pacific Ocean basin scale. Similar to the trends observed during ENSO events, sea temperature anomalies, pycnocline, thermocline, and sea level slopes are accentuated on the basin scale during PDO cool or warm regimes (e.g. Kosaka & Xie, 2013; Montecinos & Pizarro, 2005; Pizarro & Montecinos, 2004; Salinger, Renwick, & Mullan, 2001). During cool regimes, these slopes are lower in the eastern Pacific and higher in the western Pacific, whereas the opposite occurs during warm regimes. This interdecadal large-scale pattern is driven by equatorial wind variability and an atmospheric teleconnection with the South Pacific Anticyclone. On average, from 1960 to 2014 off Peru, sea-surface temperature anomalies of ±0.5 °C were evidenced on this timescale, associated with a deepening/shoaling of the thermocline of 5–10 m during warm/cold regimes (Grados et al., 2018). Interdecadal variability of warm and cool PDO regimes superimposes on interannual variability of warm and cool ENSO modulating the vertical mixing and primary production in the HCS. A regime shift in the structure and function of the entire Pacific ecosystem was discovered associated with this interdecadal variability (Chavez, Ryan, Lluch-Cota, & Ñiquen, 2003). All over the Pacific Ocean, several changes in composition and abundances of marine organism have been found, such as the alternation of a cool anchovy regime and a warm sardine regime evidenced during the 1970s. In tropical and subtropical EBUS such as the northern HCS, a shallow thermocline is associated with strong upwelling process during the cool PDO regime. The high nutrient transport enhanced the primary production which consequently increases the zooplankton and anchovy abundance. Conversely, during warm PDO regime, a deeper thermocline is associated with weakened upwelling process, reduced nutrient transport and primary production. That low primary production is associated with a decrease in the zooplankton and anchovy abundances, the latter being replaced by an increase in sardine abundance. Moreover, a significant cooling of the eastern tropical Pacific sea-surface temperatures (SST) has been evidenced during the 1979–2014 period, associated with cool PDO regime (e.g. Clem, Renwick, & McGregor, 2017; England et al., 2014; Falvey & Garreaud, 2009). Since the 2000s, a recent slowdown in the rate of global warming has even been partially attributed to this cooling (e.g. England et al., 2014; Kosaka & Xie, 2013).

In the past 60 years, the overall HCS has shown alongshore wind and upwelling intensifications (reviewed in last IPCC report on the ocean and cryosphere by Pörtner et al., 2019). However, there are still ongoing debates on future trends of the upwelling process along the coast off Peru and Chile, and projections should be taken with caution. Recent studies support that, in the summertime, upwelling-favorable winds weaken off the Peruvian coast (Goubanova et al., 2011; Rykaczewski et al., 2015). On the contrary, on the southern tip of the HCS, Rykaczewski et al. (2015) predicted that wind stress will intensify during summer and Goubanova et al. (2011) found that wind stress already experiences a significant intensification, but during Austral winter. These large-scale trends may affect considerably the primary production, which peaks during the spring-summer period all along the coast of the HCS. Oyarzún and Brierley (2019) also concluded that wind stress will increase off the Chilean coast during the twenty-first century, with a stronger trend in the southern region. As a consequence of increasing wind stress, an enhanced coastal upwelling should be observed on the upper layer. However, upwelling below 100 m depth should be reduced by increasing ocean stratification. If such predictions are confirmed, marine ecosystems from the upper layer would be largely deprived of upwelled rich-nutrient flows. In addition, a poleward shift of the South Pacific Anticyclone (SPA) has been evidenced since 2007 (Aguirre, García-Loyola, Testa, Silva, & Farias, 2018; Schneider, Donoso, Garcés-Vargas, & Escribano, 2017) and may be attributed to global climate change. In south-central Chile (~35–42°S), the displacement of the SPA considerably intensifies the upwelling-favorable winds, and particularly in winter. Consequently, the enhanced upwelling process in the southern tip of the HCS is associated in the upper layer with SST cooling, increases in primary productivity and salinity, along with a potential decline in dissolved oxygen concentration. Furthermore, Cai et al. (2018) recently reached the first inter-model consensus on future variability of the ENSO phenomenon under greenhouse warming. Associated with enhanced stratification in the upper equatorial Pacific, we should expect more extreme Eastern-Pacific El Niño events as well as extreme La Niña events (Cai et al., 2015), and consequently, strong impacts on the HCS biophysical features and ecosystems. The HCS has also registered strongly increasing trends in deoxygenation in the last few decades (Pörtner et al., 2019). As for the California EBUS, the expanding minimum oxygen zone may significantly alter the ecosystem structure and fisheries catches within the HCS, although the direction and magnitude of observed changes tend to vary among and within EBUS. Moreover, as in any other marine regions of the world, it is certain that the HCS is facing acidification (e.g. Hoegh-Guldberg & Bruno, 2010), and there is high confidence that the HCS will experiment calcium carbonate and aragonite undersaturation within decades (Pörtner et al., 2019). Combined with decreasing oxygen levels, it will increasingly affect shellfish larvae, benthic invertebrates, demersal fishes, and associated fisheries and aquaculture. The high biophysical variability of the HCS, along with uncertainties in future projections of its main features, makes the predictionsv of the effects of climate change on marine ecosystems and fisheries particularly challenging. Nevertheless, there is no doubt that it will have very strong and heterogenous impacts within the HCS.

1.2 Diversity, Productivity, and Trends in Fisheries Landings

The HCLME, in its distinct biogeographic units, is habitat for a rich and productive biota. Camus (2001), based on an extensive review of biogeographic classifications, divided the HCLME into three spatial units: the Peruvian Province (north of 30°S) containing a warm temperate biota; the Magellanic Province (41–43°S to 56°S) with an austral biota; and an extensive Intermediate area (30°S to 41°–43°S) containing a mixed biota without a distinguishing character. Species richness assessment was undertaken by Miloslavich et al. (2011) for the entire HCLME. It recorded 10.201 species, including 77 introduced species where Crustacea (N = 3136), Mollusca (N = 1203), and Pisces (N = 1167) were the three most diverse taxonomic groups. Only 1.5% of this richness is currently used as resources by fisheries. However, in this total area of 2.5 million km^2, which represents less than 0.01% of the world oceans (361 million km^2), fishery landings represented up to 25% (13.7 Mt/year) of world fishery landings for the most productive years (Fig. 2). Over the last five years, landings moved between 7 and 10% of the global landings (6–10 Mt/year) (Fig. 2a). Peru landed up to 12 Mt in 1970, with a second peak of 11.9 Mt in 1994. Chile had its highest landings of up to 7.8 Mt in the 1990s (Fig. 2b). At present Peru lands 3–7 Mt and Chile 1–3 Mt.

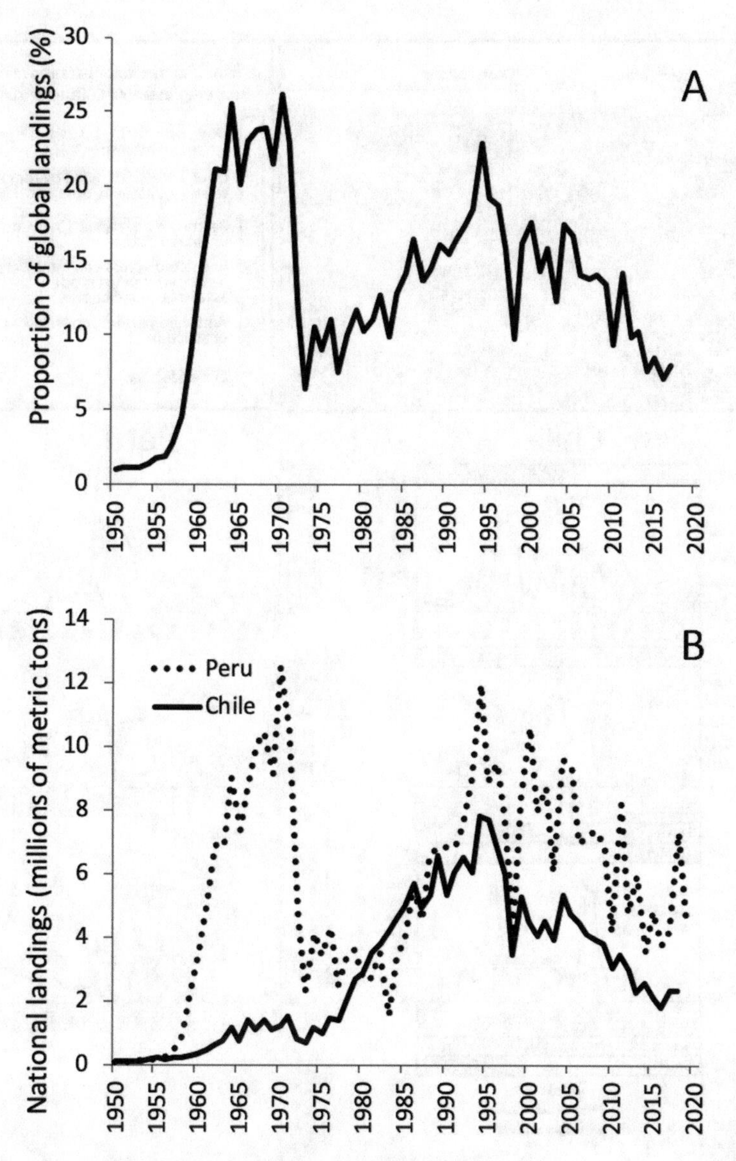

Fig. 2 Fishery landings of the Humboldt Current Large Marine Ecosystem (HCLME). (**a**) Proportion of world marine fishery landings; (**b**) Total landings in Peru and Chile

Among the various species used as resources by fisheries, high landings are supported by only a few groups (Fig. 3a). The most heavily fished are the neritic and oceanic pelagic fishes, with landings in the hundreds of thousands to millions of tons (Fig. 3a). These are mainly five species, the jack and chub mackerels (*Trachurus murphyi* and *Scomber japonicus*), and the small grazers, anchovy, and sardine (*Engraulis ringens* and among the sardines mainly *Strangomera bentincki* and

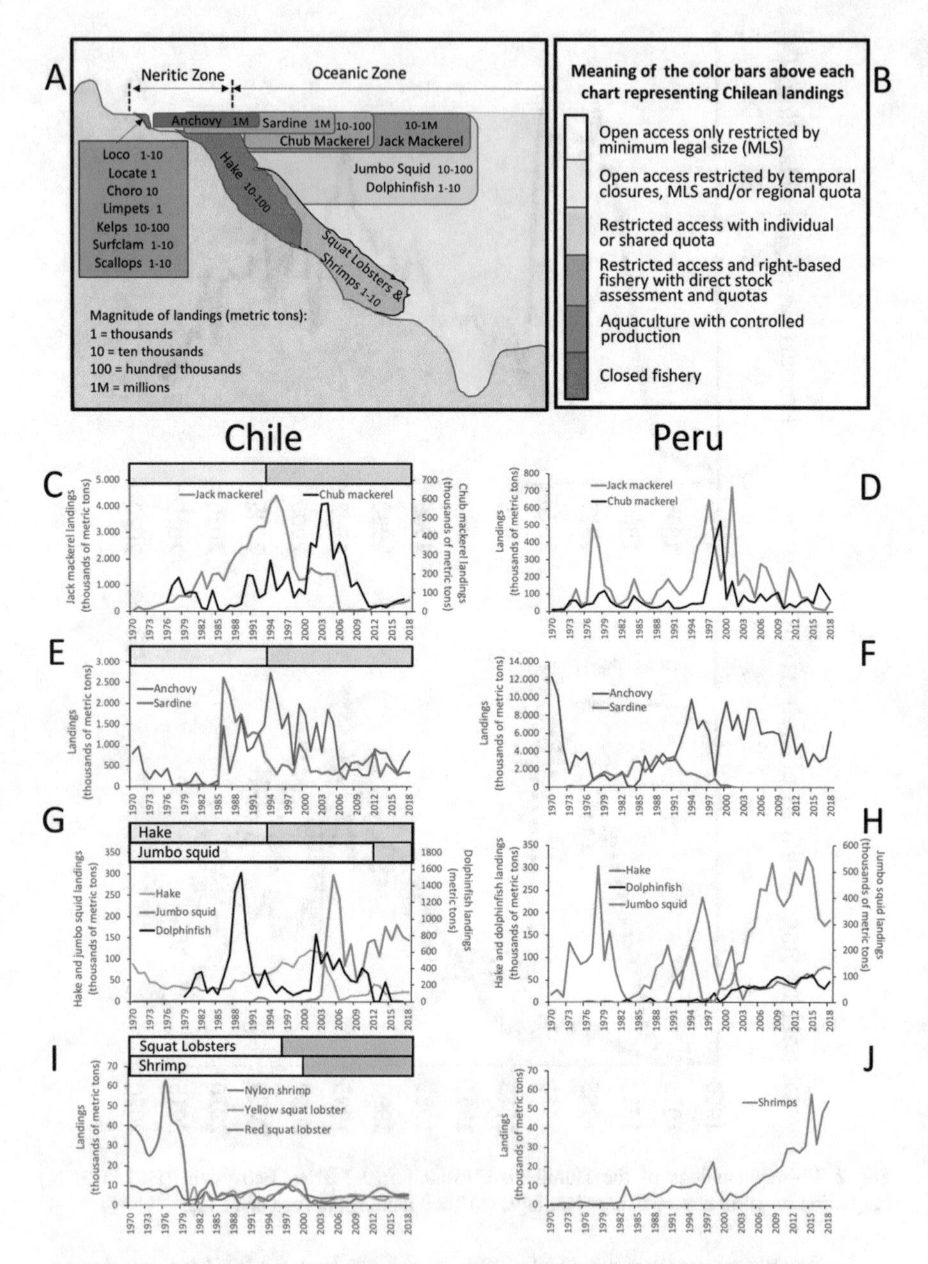

Fig. 3 Landings of pelagic and demersal fisheries in Peru and Chile. (**a**) General distribution of fish stocks in the HCLME (adapted from Tarazona, Gutiérrez, Paredes, & Indacochea, 2003). The magnitude of landings of each species is indicated. (**b**) Different management regimes applied to Chilean fisheries. This is the legend that corresponds to the horizontal color bars above each graphic representing Chilean landings in Fig. 4c, e, g, i. (c–j) Landings of diverse pelagic and demersal species. Left: Chilean fisheries landings; Right: Peruvian fisheries landings

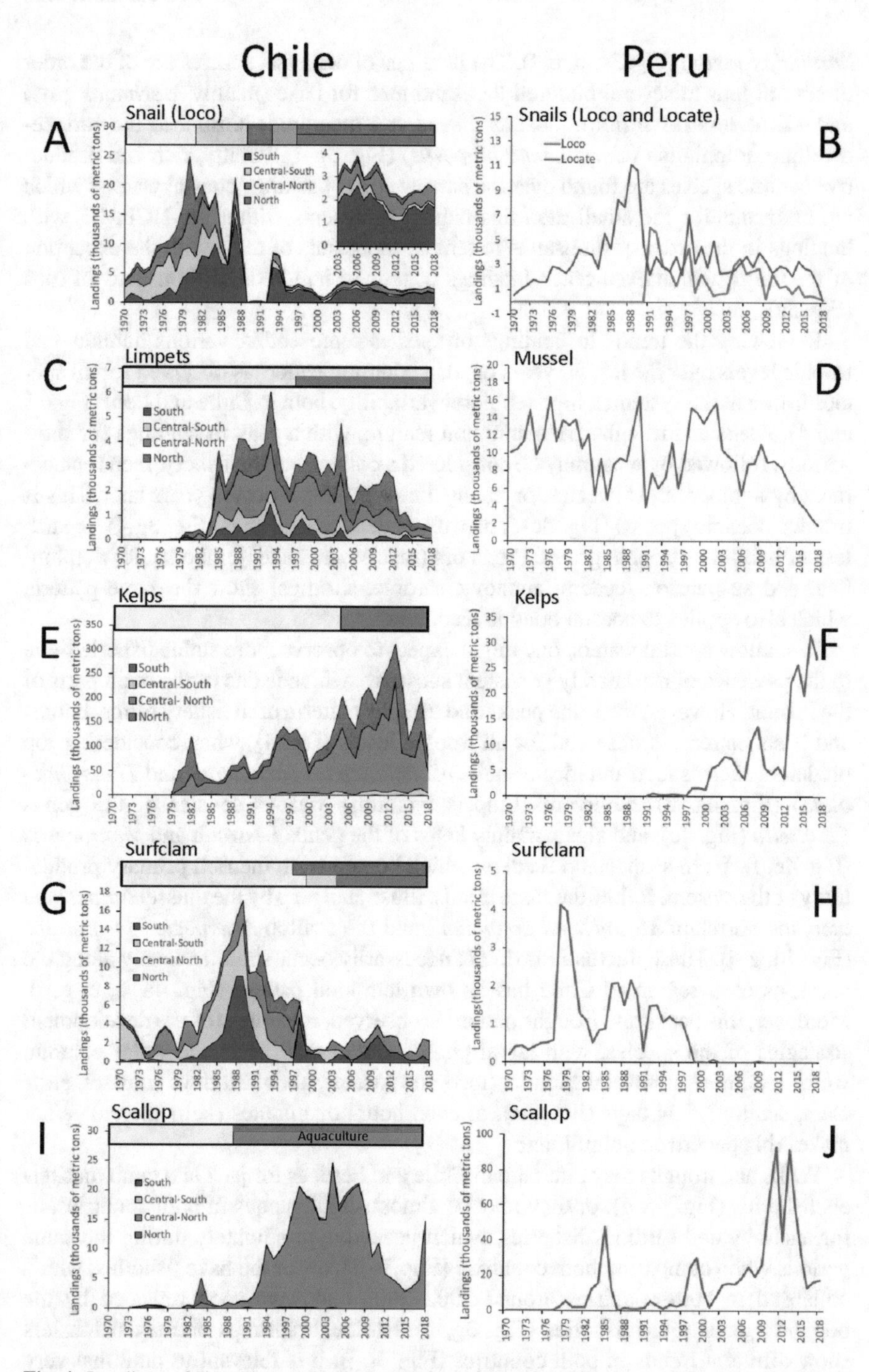

Fig. 4 Landings of diverse coastal benthic species in Peru and Chile. Left: Chilean fisheries landings; Right: Peruvian fisheries landings. See Fig. 3b in order to obtain the meaning of the color bars above each chart representing Chilean landings (Fig. 4a, c, e, g, i)

Sardinops sagax) (Fig. 3c, d, e, f). The landings of demersal species are of the order of several tens to several hundred thousand tons for hake (mainly *Merlucius gayi*) and a little less for shrimp, lobsters, fished at a magnitude similar to the surface-dwelling dolphinfish (*Coryphaena hippurus*) (Fig. 3g–j). Finally, rich and productive benthic species are found over the narrow strip of shallow coastal waters, which are notorious for the small area this habitat represents within the HCLME, with landings in the order of thousands to tens of thousands of tons, with the exception of the kelps, which even reach landings of tens to hundreds of thousands of tons (Fig. 3a).

Reviewing the trends in landings of species representing various habitats and trophic levels over the last 50 years or so, a common pattern is observed for all species fished in the system: a high temporal variability, both in Chile and Peru (Figs. 3 and 4). Peaks and troughs are a common feature, with a peak in landings for short periods, followed by what may be considered a collapse of the fishery, then replacement by another target species, or finally a new peak that occurs years later. This is true for oceanic species (Fig. 3c–f) as well as demersal species (Fig. 3g–j), regardless of habitat or trophic group. Predators (mackerels, hake, jumbo squid, dolphinfish) and suspension feeders (anchovy, sardine, shrimps) show the same pattern, which also applies to coastal benthic resources.

In shallow coastal waters, one might expect to observe more stable fisheries, due to the presence of a relatively persistent substrate, which is one of the main parts of the habitat. However, the same peaks and troughs pattern, or in fishery terms, booms and busts pattern, is observed for all trophic levels (Fig. 4), when considering top predators such as loco and locate snails (*Concholepas concholepas* and *Thais chocolata*) (Fig. 4a, b), herbivores limpets including various species of the genus *Fissurella* (Fig. 4c), and algae, mainly kelps of the genus *Lessonia* and *Macrocystis* (Fig. 4e, f). Even suspension feeders, which benefit from the rich primary productivity of the system, follow the same trends, illustrated here by the mussel *Aulacomya ater*, the surfclam *Mesodesma donacium,* and the scallop *Argopecten purpuratus* (Fig. 4d, g–j). These fluctuations do not necessarily occur simultaneously along the coast, as each region of Chile has its own temporal pattern (Fig. 4a, c, e, g, i). Moreover, this peaks and troughs pattern is observed, regardless of the development strategies of the species, with larval phases and/or drifting in the water column, which extend over several months (loco and locate), about 30 days (mussel, surfclam, scallop), few days (limpets), or even hours or minutes (kelp spores). What makes this pattern so ubiquitous?

Peaks and troughs may alternate in Chile and Peru, as for jack and chub mackerels fisheries (Fig. 3c, d), or they may be almost simultaneous in both countries, as for anchovy and sardines fisheries, which peaked approximately during the same periods when comparing both countries (Fig. 3e, f), or for the hake fisheries, which collapsed in both countries around 2002–2003 and were soon replaced by the booming giant squid fisheries (Fig. 3g, h). Although shrimps and squat lobsters show different trends in both countries (Fig. 3i, j), it is relevant to note that very distinct species are fished along the Chilean and Peruvian coasts. While in Chile, the main species are the squat lobsters *Pleuroncodes monodon* and *Cervimunida johni*, and the Chilean nylon shrimp *Heterocarpus reedi*, in Peru, diverse Penaeid shrimps are caught, including *Litopenaeus vannamei*, *L. stylirostris*, *L. occidentalis*, *Farfantepenaeus californiensis*, *F. brevirostris*, *Rimapenaeus fuscina,* and *Xiphopenaeus riveti* according to Ordinola Zapata et al. (2008).

When a peak in landings is followed by a steep decline, the most common interpretation is that it occurs as a result of overfishing and a lack of adequate management. Nevertheless, with different and separate fleets, fisheries administrations and management strategies established in Chile and Peru, we observe similar general patterns when comparing landings of the same species. Thus, this argument of lack of adequate management is questionable, to say the least. Moreover, when observing in detail the implementation of new management measures to regulate landings from Chilean fisheries, we see that increasingly strict control measures are observed (cf. the grey tone in the bar above each graph, which becomes darker or even red as the fisheries are closed (Fig. 3c, e, g, i and Fig. 4a, c, e, g, i, according to the legend shown in Fig. 3b). Although the management regime is strengthened over time for all species analyzed, little change is observed in the peaks and troughs pattern. Indeed, stricter control for some species reduces the magnitude of landings, but zooming in on the landings time series (e.g. Fig. 4a), and the same pattern of peaks and troughs is again observed, only at a lower magnitude. This suggests that the HCLME fisheries benefit from a highly productive but very variable source of production, over which management has little influence.

According to extensive literature, fishery yields respond primarily to a combination of environmental and anthropogenic stressors at several scales (e.g. Hofmann & Powell, 1998; Crona et al., 2015; Yáñez et al., 2017; Belhabib, Dridi, Padilla, Ang, & Le Billon, 2018). Within the HCLME, there is strong evidence of environmental forcing occurring on pelagic fisheries landings (e.g. Arcos, Cubillos, & Núnez, 2001; Fréon, Cury, Shannon, & Roy, 2005; Plaza, Salas, & Yáñez, 2018; Yáñez, Barbieri, Silva, Nieto, & Espındola, 2001), of which the most remarkable demonstration may be the major role of the extreme El Niño event of 1972–1973 in the collapse of the Peruvian anchovy fishery, along with undeniable overfishing effect (e.g. Pauly et al., 2002) enhanced by an interdecadal ecosystem shift from an anchovy to a sardines' regime (e.g. Chavez et al., 2003). In such context, the likely influence of climate change is of particular concern (e.g. Yáñez, Lagos, et al. 2017, Yáñez, Plaza, et al. 2017; Silva, Leiva, & Lastra, 2019; Silva, Yáñez, Barbieri, Bernal, & Aranis, 2015). In comparison, few studies have been carried out on the influence of biophysical variability on coastal benthic fisheries in the HCLME (e.g. Defeo et al., 2013). However, its documented effects on benthic communities, populations recruitment, and physiological traits (e.g. Barahona et al., 2019; Flores-Valiente et al., 2019; Ramajo et al., 2019; Vásquez, Vega, & Buschmann, 2006) suggest that environmental variability may highly contribute to the peaks and troughs pattern observed in benthic fisheries. As with pelagic fisheries, climate change effects on the coastal benthic ecosystem of the HCLME are of prior concern for benthic fisheries, in particular ocean acidification and increasing temperature effects on species physiological traits (e.g. Diaz, Lardies, Tapia, Tarifeño, & Vargas,

2018; Navarro et al., 2016), as well as species-specific adaptive responses to this stressors (e.g. Vargas et al., 2017).

As in other harvested ecosystems of the world, it therefore appears that effective fishery management requires a constant monitoring and flexible adaptation to this environmental variability, as well as considering its own uncertainty (e.g. Carpenter, Brock, Folke, van Nes, & Scheffer, 2015; Polasky, Carpenter, Folke, & Keeler, 2011; Schindler & Hilborn, 2015). The little influence that management has on fishery landings fluctuations, as illustrated along the Chilean coast, appears to be largely attributed to its difficulty coping with high spatial and temporal environmental variability in the HCLME, which led to the collapse of certain fisheries (e.g. Aburto & Stotz, 2013; Stotz, 2018; Thiel et al., 2007). Even the most stringent management, such as the one that was implemented in the case of the loco fishery (*Concholepas concholepas*), after a perceived overfishing crisis in the 1980s, which led to a closure of the loco fishery (1989–1992), did not attenuate much of the variability (Fig. 4c). Since 1997, the loco fishery has been administered through the allocation of exclusive fishing rights (or privileges) to fishermen's organizations on designated coastal areas of tens to hundreds of hectares (Castilla & Gelcich, 2008; Stotz, 1997). Outside these areas, the fishery has remained closed (Fig. 4c—half of the bar appears in red). The assigned areas are managed by direct stock assessments, on the basis of which catch quotas are estimated and authorized each year. Management is carried out by the fishermen, who, according to the regulations, must be assisted by a marine science professional (marine biologist, fisheries or aquaculture engineer, etc.) and supervised by the State authority, which reviews reports and authorizes the quotas. This strict management regime has effectively led to a reduction in national legal landings, but without reducing the variability of catch (Fig. 4c). Indeed, landings from individual areas vary greatly along the coast and over time, leaving even fishermen without a catch in some years (Stotz, 2018; Stotz, Arias, & Miranda, 2019). This management regime of Territorial Use Rights in Fisheries (TURFs), originally implemented for the loco fisheries, was quickly extended to many other benthic resources (other gastropods, algae, bivalves, cephalopods, crustaceans, equinoderms, and a tunicate). While mixed results have been attributed to TURF management in Chile (Gelcich et al., 2017; Stotz, 2018), it substantially contributed to settle down the benthic small-scale fishermen along the Chilean coast (Aburto, Thiel, & Stotz, 2009). Deprived of a migratory behavior, which was shaped by the high temporal and spatial variability in the HCLME, fishermen have faced new difficulties (Stotz, 2018), and the State has failed to solve them through subsidies and/or investments because it has not adequately taken into account fishermen idiosyncrasies (Stotz et al., 2019). Another case study is the development of scallop aquaculture (*Argopecten purpuratus*) in Chile (Fig. 4i), which has reached a historical production threshold five times higher than the production threshold of the scallop fishery, without however allowing a stabilization of the production of this resource. Most of their lives, these cultivated scallops are exposed to environmental high variability, while growing up at sea (~12–15 months) along lantern-nets (Von Brand, Abarca, Merino, & Stotz, 2016), which consist of vertical alignments of cylindrical cultivation compartments. The favorable effect of

1982–1983 and 1997–1998 extreme El Niño events on the recruitment and growth of *Argopecten purpuratus* is well-documented in central and southern Peru (Mendo, Wolff, Mendo, & Ysla, 2016), and quite consistent with high landings and a peak in aquaculture production observed for the following years in northern Chilean and Peruvian fisheries (Fig. 4i, j). In addition, as for the capture fisheries, a large part of the fluctuations in scallop production depends on the volatility of the market, the management system implemented and the social interactions between the different actors involved, in interaction with the environmental variability observed. These are the critical aspects we are going to develop now.

Fishery management policies have to be understood within a wider political context. Since the 1980s, in a context of transition towards neoliberal policies in Latin America, a strong emphasis on private property rights, economic efficiency, government budget cuts, and delegation of responsibilities and risks to the private sector have become increasingly prevalent in most of the Latin American states, as has been the case in Chile, spurred on by military dictatorship, and even well established in Peru, where political populism coexisted with neoliberalism during the 1990–2000 period (Roberts, 1995; Weyland, 2001, 2004). This short piece of history must be carefully contextualized within a longer path, built around natural resources massive exploitation and open access management. In this neoliberal context, by taking up Polanyi's (2001) statement on capitalism, both labor and the biophysical environment have been subjected to commodification, as to satisfy the international market demand. Therefore, as in most parts of the world, fisheries of the HCLME have had to confront and adapt to a market-based approach, export-oriented, along with privatization and trade liberalization (e.g. Campling, Havice, & McCall Howard, 2012; Ibarra, Reid, & Thorpe, 2000; Pinkerton & Davis, 2015). Campling et al. (2012) reviewed some of the consequences of this market-based approach on world fisheries, where fishers struggle to remain competitive with distant rivals facing other fishing conditions, collaborate with leading actors of the market chains often neglecting essential relationships with local actors, are exposed to sudden change in target species and price relations, risk to deteriorate environmental conditions that sustain their livelihoods, and may tend to evade regulation through illegal practices. The development of the Chilean and Peruvian scallop aquaculture, and more generally the peaks and troughs pattern observed in production of coastal benthic and pelagic resources, have to be analyzed and understood within this scenario.

The scallop was traditional and locally consumed seafood before the 1980s in Chile and Peru. El Niño event of 1982–1983 changed this dramatically. As previously stated, a huge recruitment occurred in northern Chile (Antofagasta Region) and southern Peru (Bahía Independencia), producing a notorious but brief peak of landings during those years (Fig. 4i and j). In a favorable neoliberal context, the abundance of scallops at that time opened Chilean and Peruvian fisheries to international markets (Mendo et al., 2016), providing a crucial incentive for subsequent development. In Chile, natural beds collapsed, as abundance rapidly declined, while the fishery maintained its efforts to satisfy the export market. A fishing ban has been established since 1986 and has so far been maintained in the country. Therefore, the only way to produce scallops in Chile has become aquaculture, which started in 1984. As a result of the fishing ban and the strong demand on the export market, scallop aquaculture highly developed during the 1990s (Von Brand et al., 2016). However, because aquaculture was also an incentive for illegal fishing through its demand for "seeds" (i.e. juvenile organisms), natural beds almost disappeared, and the scallop has become almost a "domestic" species in Chile, with production of a much higher order of magnitude than landings from the former fishery (Fig. 4i) (Stotz, 2000). In Peru, in the meantime, the fishery returned to the same levels as before the 1982 ENSO event, until the 1997 ENSO event produced a new peak in southern Peru (Fig. 4j). After the fishery collapsed in this region, also linked to poor management, many fishers migrated northward, to the Sechura Bay (Kluger, Taylor, Wolff, Stotz, & Mendo, 2019), exploiting natural beds that had developed in northern Peru in the inter-period between the two extreme ENSO events, which adversely affected scallop populations in this region (Mendo et al., 2016). With the development of low-cost and highly productive bottom cultures in the Sechura Bay, scallop production in Peru rapidly grew, with nearly 100.000 tons produced in 2010 (Fig. 4j). Since then, Peru has become the leading producer in Latin America. After a boom in Chilean aquaculture between 1990 and 2007, production dropped in Chile in 2007–2009, due to falling prices triggered by massive Peruvian production (Fig. 4i, j). Moreover, less favorable environmental and socioeconomic conditions did not allow Chilean producers to compete with Peruvian production methods and the efficient societal model organized around scallop production (Kluger et al., 2019). Finally, the 2010 earthquake in Chile, followed by the aftershocks of the 2011 tsunami in Japan completed weakening the Chilean production, as most fishing enterprises ceased operations due to heavy losses of cultivation materials and scallops (Von Brand et al., 2016). However, the low natural seed supply observed in northern Peru since 2015 (Kluger et al., 2019) could be favorable to the remaining Chilean producers, and competition in the export market may be back on track, along with deregulated market and high biophysical variability.

2 Conclusion

The large fluctuations observed in *Argopecten purpuratus* scallop production in the HCLME illustrate to the extreme, the close interaction between environmental and market variabilities, along with the implementation of management measures within a dynamic social and political context. To a similar extent, the landings of each one of the pelagic and coastal benthic resources of the HCLME are embedded in such complex social-ecological systems (SESs) (e.g. Ostrom, 2009). The greatest challenge for modelers is to identify each one of the key variables in these fisheries systems and understand how they interact on different spatial and temporal scales. Considering this complexity should help managers and fishers to learn how to take advantage of it, rather than pretending and failing to eliminate it from such systems. This means that flexible and adaptive regulations need to be implemented to be able to respond to short term variability, accepting the large amount of uncertainty inherent in a dynamic and productive system. Past successes and failures of management actions, extreme events, fluctuations of correlated variables across space and time, feedback loops, non-linearity patterns, and successions of order and chaos must be carefully tracked and analyzed, contributing to the understanding needed to address current issues. While the HCLME is one of the most productive marine ecosystems on Earth, along with a great biophysical and biological variability, it is also a huge but fragile exploited social-ecological system. Modelers can help to understand how to continuously adapt in order to deal with recurrent fisheries "crises," which, as we are beginning to learn, may simply be an expression of the variability inherent in this complex system. Furthermore, it is essential to increase our capacity to adapt to the expected consequences of climate change.

References

Aburto, J., & Stotz, W. (2013). Learning about TURFs and natural variability: Failure of surf clam management in Chile. *Ocean & Coastal Management, 71*, 88–98.

Aburto, J., Thiel, M., & Stotz, W. (2009). Allocation of effort in artisanal fisheries: The importance of migration and temporary fishing camps. *Ocean & Coastal Management, 52*(12), 646–654.

Aguirre, C., García-Loyola, S., Testa, G., Silva, D., & Farias, L. (2018). Insight into anthropogenic forcing on coastal upwelling off south-central Chile. *Elementa Science of the Anthropocene, 6*(59), 1–13.

Allison, E. H., Delaporte, A., & Hellebrandt de Silva, D. (2013). *Integrating fisheries management and aquaculture development with food security and livelihoods for the poor. Report submitted to the Rockefeller Foundation* (p. 124). Norwich: School of International Development, University of East Anglia, Norwich, United Kingdom.

Arcos, D. F., Cubillos, L. A., & Núnez, S. P. (2001). The jack mackerel fishery and El Niño 1997–98 effects off Chile. *Progress in Oceanography, 49*(1–4), 597–617.

Arntz, W., & Fahrbach, E. (1996). *El Niño, experimento climático de la naturaleza* (p. 312). México: Fondo de Cultura Económica.

Barahona, M., Broitman, B. R., Faugeron, S., Jaugeon, L., Ospina-Alvarez, A., Véliz, D., & Navarrete, S. A. (2019). Environmental and demographic factors influence the spatial genetic structure of an intertidal barnacle in central-northern Chile. *Marine Ecology Progress Series, 612*, 151–165.

Barange, M., Bahri, T., Beveridge, M. C. M., Cochrane, K. L., Funge-Smith, S., & Poulain, F., (eds.) (2018). Impacts of climate change on fisheries and aquaculture: Synthesis of current knowledge, adaptation and mitigation options. *FAO Fisheries and Aquaculture Technical Paper* No. 627. Rome, p. 628.

Behrenfeld, M. J., O'Malley, R. T., Siegel, D. A., McClain, C. R., Sarmiento, J. L., Feldman, G. C., Milligan, A. J., Falkowski, P. G., Letelier, R. M., & Boss, E. S. (2006). Climate-driven trends in contemporary ocean productivity. *Nature, 444*(7120), 752–755.

Belhabib, D., Dridi, R., Padilla, A., Ang, M., & Le Billon, P. (2018). Impacts of anthropogenic and natural "extreme events" on global fisheries. *Fish and Fisheries, 19*(6), 1092–1109.

Bertrand, A., Vögler, R., & Defeo, O. (2018). Climate change impacts, vulnerabilities and adaptations: Southwest Atlantic and Southeast Pacific marine fisheries. In Barange, M., Bahri, T., Beveridge, M. C. M., Cochrane, K. L., Funge-Smith, S., & Poulain, F. (eds.). Impacts of climate change on fisheries and aquaculture: Synthesis of current knowledge, adaptation and mitigation options. *FAO Fisheries and Aquaculture Technical Paper* No. 627, Rome, pp. 325–346.

Bonhomme, C., Aumont, O., & Echevin, V. (2007). Advective transport caused by intraseasonal Rossby waves: A key player of the high chlorophyll variability off the Peru upwelling region. *Journal of Geophysical Research: Oceans, 112*(9), 18–30.

Bravo, L., Ramos, M., Astudillo, O., Dewitte, B., & Goubanova, K. (2016). Seasonal variability of the Ekman transport and pumping in the upwelling system off central-northern Chile (~30° S) based on a high-resolution atmospheric regional model (WRF). *Ocean Science, 12*(5), 1049–1065.

Cai, W., Wang, G., Dewitte, B., Wu, L., Santoso, A., Takahashi, K., Yang, Y., Carréric, A., & McPhaden, M. J. (2018). Increased variability of eastern Pacific El Niño under greenhouse warming. *Nature, 564*(7735), 201–206.

Cai, W., Wang, G., Santoso, A., McPhaden, M. J., Wu, L., Jin, F., Timmermann, A., Collins, M., Vecchi, G., Lengaigne, M., England, M. H., Dommenget, D., Takahashi, K., & Guilyardi, E. (2015). Increased frequency of extreme La Niña events under greenhouse warming. *Nature Climate Change, 5*(2), 132–137.

Campling, L., Havice, E., & McCall Howard, P. (2012). The political economy and ecology of capture fisheries: market dynamics, resource access and relations of exploitation and resistance. *Journal of Agrarian Change, 12*(2–3), 177–203.

Camus, P. A. (2001). Biogeografía marina de Chile continental. *Revista Chilena De Historia Natural, 74*(3), 587–617.

Carpenter, S. R., Brock, W. A., Folke, C., van Nes, E. H., & Scheffer, M. (2015). Allowing variance may enlarge the safe operating space for exploited ecosystems. *Proceedings of the National Academy of Sciences, 112*(46), 14384–14389.

Carr, M. E., & Kearns, E. J. (2003). Production regimes in four Eastern boundary current systems. *Deep Sea Research Part II: Topical Studies in Oceanography, 50*(22–26), 3199–3221.

Castilla, J. C., & Gelcich, S. (2008). Management of the loco (Concholepas concholepas) as a driver for self-governance of small-scale benthic fisheries in Chile. In Townsend, R., R. Shotton, H. Uchida (eds). Case studies in fisheries self-governance. *FAO Fisheries Technical Paper* No. 504, Rome, pp. 441–451.

Chavez, F. P., Bertrand, A., Guevara-Carrasco, R., Soler, P., & Csirke, J. (2008). The northern Humboldt current system: Brief history, present status and a view towards the future. *Progress in Oceanography, 79*, 95–105.

Chavez, F. P., Ryan, J., Lluch-Cota, S. E., & Ñiquen, M. (2003). From anchovies to sardines and back: multidecadal change in the Pacific Ocean. *Science, 299*(5604), 217–221.

Clarke, A. J., & Shi, C. (1991). Critical frequencies at ocean boundaries. *Journal of Geophysical Research: Oceans, 96*(C6), 10731–10738.

Clem, K. R., Renwick, J. A., & McGregor, J. (2017). Relationship between eastern tropical Pacific cooling and recent trends in the Southern Hemisphere zonal-mean circulation. *Climate Dynamics, 49*(1–2), 113–129.

Costello, C., Cao, L., Gelcich, S., Cisneros, M. A., Free, C. M., Froehlich, H. E., Galarza, E., Golden, C. D., Ishimura, G., Macadam-Somer, I., Maier, J., Mangin, T., Melnychuk, M. C., Miyahara, M., de Moor, C., Naylor, R., Nostbakken, L., Ojea, E., O'Reilly, E., Osio, G. C., Parma, A. M., Pina Amargos, F., Plantinga, A. J., Tacon, A., & Thilsted, S. H. (2019). *The future of food from the Sea* (p. 60). Washington, DC: World Resources Institute.

Crona, B. I., Van Holt, T., Petersson, M., Daw, T. M., & Buchary, E. (2015). Using social–ecological syndromes to understand impacts of international seafood trade on small-scale fisheries. *Global Environmental Change, 35*, 162–175.

Defeo, O., Castrejón, M., Ortega, L., Kuhn, A. M., Gutiérrez, N. L., & Castilla, J. C. (2013). Impacts of climate variability on Latin American small-scale fisheries. *Ecology and Society, 18*(4), 30.

Demarcq, H. (2009). Trends in primary production, sea surface temperature and wind in upwelling systems (1998–2007). *Progress in Oceanography, 83*(1–4), 376–385.

Dewitte, B., Vazquez-Cuervo, J., Goubanova, K., Illig, S., Takahashi, K., Cambon, G., Purca, S., Correa, D., Gutiérrez, D., Sifeddine, A., & Ortlieb, L. (2012). Change in El Niño flavours over

1958–2008: Implications for the long-term trend of the upwelling off Peru. *Deep Sea Research Part II: Topical Studies in Oceanography, 77*, 143–156.

Diaz, R., Lardies, M. A., Tapia, F. J., Tarifeño, E., & Vargas, C. A. (2018). Transgenerational effects of pCO2-driven ocean acidification on adult mussels Mytilus chilensis modulate physiological response to multiple stressors in larvae. *Frontiers in Physiology, 9*, 1349.

Echevin, V., Albert, A., Lévy, M., Graco, M., Aumont, O., Piétri, A., & Garric, G. (2014). Intraseasonal variability of nearshore productivity in the Northern Humboldt Current System: The role of coastal trapped waves. *Continental Shelf Research, 73*, 14–30.

England, M. H., McGregor, S., Spence, P., Meehl, G. A., Timmermann, A., Cai, W., Gupta, A. S., McPhaden, M. J., Purich, A., & Santoso, A. (2014). Recent intensification of wind-driven circulation in the Pacific and the ongoing warming hiatus. *Nature Climate Change, 4*(3), 222–227.

Falvey, M., & Garreaud, R. D. (2009). Regional cooling in a warming world: Recent temperature trends in the southeast Pacific and along the west coast of subtropical South America (1979–2006). *Journal of Geophysical Research: Atmospheres, 114*(4), 102–117.

FAO. (2018). The State of World Fisheries and Aquaculture 2018—Meeting the sustainable development goals. Licence: CC BY-NC-SA 3.0 IGO, p. 227.

Figueroa, D., & Moffat, C. (2000). On the influence of topography in the induction of coastal upwelling along the Chilean coast. *Geophysical Research Letters, 27*(23), 3905–3908.

Flores-Valiente, J., Tam, J., Brochier, T., Colas, F., Pecquerie, L., Aguirre-Velarde, A., Mendo, J., & Lett, C. (2019). Larval supply of Peruvian scallop to the marine reserve of Lobos de Tierra Island: A modeling approach. *Journal of Sea Research, 144*, 142–155.

Fréon, P., Barange, M., & Aristegui, J. (2009). Eastern boundary upwelling ecosystems: integrative and comparative approaches preface. *Progress in Oceanography, 83*(1–4), 1–14.

Fréon, P., Cury, P., Shannon, L., & Roy, C. (2005). Sustainable exploitation of small pelagic fish stocks challenged by environmental and ecosystem changes: a review. *Bulletin of Marine Science, 76*(2), 385–462.

Gelcich, S., Cinner, J., Donlan, C. J., Tapia-Lewin, S., Godoy, N., & Castilla, J. C. (2017). Fishers' perceptions on the Chilean coastal TURF system after two decades: problems, benefits, and emerging needs. *Bulletin of Marine Science, 93*(1), 53–67.

Golden, C. D., Allison, E. H., Cheung, W., Dey, M., Halpern, B. S., McCauley, D. J., Smith, M., Vaitla, B., Zeller, D., & Myers, S. (2016). Nutrition: Fall in fish catch threatens human health. *Nature News, 534*(7607), 317–320.

Gomez, F. A., Spitz, Y. H., Batchelder, H. P., & Correa-Ramirez, M. A. (2017). Intraseasonal patterns in coastal plankton biomass off central Chile derived from satellite observations and a biochemical model. *Journal of Marine Systems, 174*, 106–118.

Goubanova, K., Echevin, V., Dewitte, B., Codron, F., Takahashi, K., Terray, P., & Vrac, M. (2011). Statistical downscaling of sea-surface wind over the Peru–Chile upwelling region: diagnosing the impact of climate change from the IPSL-CM4 model. *Climate Dynamics, 36*(7–8), 1365–1378.

Graco, M. I., Purca, S., Dewitte, B., Castro, C. G., Morón, O., Ledesma, J., Flores, G., & Gutiérrez, D. (2017). The OMZ and nutrient features as a signature of interannual and low-frequency variability in the Peruvian upwelling system. *Biogeosciences, 14*(46), 1–17.

Grados, C., Chaigneau, A., Echevin, V., & Dominguez, N. (2018). Upper ocean hydrology of the Northern Humboldt Current System at seasonal, interannual and interdecadal scales. *Progress in Oceanography, 165*, 123–144.

Gutiérrez, D., Grados, C., Graco, M., Vásquez, L., Velazco, F., S. Sánchez, S., Ayón, P, Tam, J., Morón, O., Flores, R., Quispe, C., & Pizarro, L. (2014). El Mar Peruano y su Dinámica. (pp. 34–59). IMARPE, Libro de Oro, Callao, Perú.

Gutierrez, D., Akester, M., & Naranjo, L. (2016). Productivity and sustainable management of the Humboldt Current large marine ecosystem under climate change. *Environmental Development, 17*, 126–144.

Helly, J. J., & Levin, L. A. (2004). Global distribution of naturally occurring marine hypoxia on continental margins. *Deep Sea Research Part I: Oceanographic Research Papers, 51*(9), 1159–1168.

Hoegh-Guldberg, O., & Bruno, J. F. (2010). The impact of climate change on the world's marine ecosystems. *Science, 328*(5985), 1523–1528.

Hoegh-Guldberg, O., Caldeira, K., Chopin, T., Gaines, S., Haugen, P., Hemer, M., Howard, J., Konar, M., Krause-Jensen, D., Lindstad, E., Lovelock, C. E., Michelin, M., Nielsen, F. G., Northrop, E., Parker, R., Roy, J., Smith, T., Some, S., & Tyedmers, P. (2019). *The ocean as a solution to climate change: Five opportunities for action* (p. 116). Washington, DC: World Resources Institute.

Hofmann, E. E., & Powell, T. M. (1998). Environmental variability effects on marine fisheries: four case histories. *Ecological Applications, 8*(sp1), S23–S32.

Hormazabal, S., Shaffer, G., & Pizarro, O. (2002). Tropical Pacific control of intraseasonal oscillations off Chile by way of oceanic and atmospheric pathways. *Geophysical Research Letters, 29*(6), 5–1.

Ibarra, A. A., Reid, C., & Thorpe, A. (2000). Neo-liberalism and its impact on overfishing and overcapitalisation in the marine fisheries of Chile, Mexico and Peru. *Food Policy, 25*(5), 599–622.

Iriarte, J. L., Vargas, C. A., Tapia, F. J., Bermúdez, R., & Urrutia, R. E. (2012). Primary production and plankton carbon biomass in a river-influenced upwelling area off Concepción, Chile. *Progress in Oceanography, 92*, 97–109.

Kämpf, J., & Chapman, P. (2016). The Peruvian-Chilean coastal upwelling system. In *Upwelling systems of the world: a scientific journey to the most productive marine ecosystems* (pp. 161–194). Berlin: Springer.

Kawarazuka, N., & Béné, C. (2010). Linking small-scale fisheries and aquaculture to household nutritional security: An overview. *Food Security, 2*(4), 343–357.

Kluger, L. C., Taylor, M. H., Wolff, M., Stotz, W., & Mendo, J. (2019). From an open-access fishery to a regulated aquaculture business: the case of the most important Latin American bay scallop (*Argopecten purpuratus*). *Reviews in Aquaculture, 11*(1), 187–203.

Kosaka, Y., & Xie, S. P. (2013). Recent global-warming hiatus tied to equatorial Pacific surface cooling. *Nature, 501*(7467), 403–407.

Lara, C., Saldías, G. S., Cazelles, B., Rivadeneira, M. M., Haye, P. A., & Broitman, B. R. (2019). Coastal biophysical processes and the biogeography of rocky intertidal species along the south-eastern Pacific. *Journal of Biogeography, 46*(2), 420–431.

Masson-Delmotte, T. W. V., Zhai, P., Pörtner, H. O., Roberts, D., Skea, J., Shukla, P. R., et al. (eds.) (2018). IPCC 2018: Summary for policymakers. In *Global warming of 1.5 C. An IPCC special report on the impacts of global warming of 1.5 C above pre-industrial levels and related global greenhouse gas emission pathways, in the context of strengthening the global response to the threat of climate change, sustainable development, and efforts to eradicate poverty* (pp. 1–24). Geneva: IPCC

McGillicuddy, D. J., Jr. (2016). Mechanisms of physical-biological-biogeochemical interaction at the oceanic mesoscale. *Annual Review of Marine Science, 8*, 125–159.

Mendo, J., Caille, G., Massutí, E., Punzón, A., Tam, J., Villasante, S., & Gutiérrez, D. 2020. Fishing Resources. In: Adaptation to Climate Change Risks in Ibero-American Countries — RIOCCADAPT Report. [Moreno, J.M., C. Laguna-Defi or, V. Barros, E. Calvo Buendía, J.A. Marengo, and U. Oswald Spring (eds.)], McGraw Hill, Madrid, Spain (ISBN 9788448621667).

Mendo, J., Wolff, M., Mendo, L., & Ysla, L. (2016). Scallop fishery and culture in Peru. In S. E. Shumway & G. J. Parsons (Eds.), *Scallops: Biology, ecology, aquaculture, and fisheries. Developments in aquaculture and fisheries science* (Vol. 40, pp. 1089–1109). Amsterdam: Elsevier.

Messié, M., & Chavez, F. P. (2015). Seasonal regulation of primary production in eastern boundary upwelling systems. *Progress in Oceanography, 134*, 1–18.

Miloslavich, P., Klein, E., Díaz, J. M., Hernandez, C. E., Bigatti, G., Campos, L., Artigas, F., Castillo, J., Penchaszadeh, P. E., Neill, P. E., Carranza, A., Retana, M. V., Díaz de Astarloa,

J. M., Lewis, M., Yorio, P., Piriz, M. L., Rodríguez, D., Yoneshigue-Valentin, Y., Gamboa, L., & Martín, A. (2011). Marine biodiversity in the Atlantic and Pacific coasts of South America: knowledge and gaps. PloS one, 6(1), e14631.

Montecinos, A., & Gomez, F. (2010). ENSO modulation of the upwelling season off southern-central Chile. *Geophysical Research Letters, 37*(2), 708–711.

Montecinos, A., & Pizarro, O. (2005). Interdecadal sea surface temperature–sea level pressure coupled variability in the South Pacific Ocean. *Journal of Geophysical Research: Oceans, 110*(8), 5–15.

Montes, I., Dewitte, B., Gutknecht, E., Paulmier, A., Dadou, I., Oschlies, A., & Garçon, V. (2014). High-resolution modeling of the Eastern Tropical Pacific oxygen minimum zone: Sensitivity to the tropical oceanic circulation. *Journal of Geophysical Research: Oceans, 119*(8), 5515–5532.

Mooers, C. N., Collins, C. A., & Smith, R. L. (1976). The dynamic structure of the frontal zone in the coastal upwelling region off Oregon. *Journal of Physical Oceanography, 6*(1), 3–21.

Navarro, J. M., Duarte, C., Manríquez, P. H., Lardies, M. A., Torres, R., Acuña, K., Vargas, C. A., & Lagos, N. A. (2016). Ocean warming and elevated carbon dioxide: multiple stressor impacts on juvenile mussels from southern Chile. *ICES Journal of Marine Science, 73*(3), 764–771.

Ordinola Zapata, E., Inga Barreto, C., & Alemán Mejía, S. (2008). Un estudio sobre langostinos (Penaeoidea) en caleta La Cruz, Tumbes. Febrero–junio 2003. Informe. Instituto del Mar del Perú, ISSN 0378-7702. 35(3), 231–240.

Ostrom, E. (2009). A general framework for analyzing sustainability of social-ecological systems. *Science, 325*(5939), 419–422.

Oyarzún, D., & Brierley, C. M. (2019). The future of coastal upwelling in the Humboldt current from model projections. *Climate Dynamics, 52*(1–2), 599–615.

Pauly, D., Christensen, V., Guénette, S., Pitcher, T. J., Sumaila, U. R., Walters, C. J., Watson, R., & Zeller, D. (2002). Towards sustainability in world fisheries. *Nature, 418*(6898), 689–695.

Pinkerton, E., & Davis, R. (2015). Neoliberalism and the politics of enclosure in North American small-scale fisheries. *Marine Policy, 61*, 303–312.

Pinochet, A., Garcés-Vargas, J., Lara, C., & Olguín, F. (2019). Seasonal variability of upwelling off Central-Southern Chile. *Remote Sensing, 11*(15), 1737–1750.

Pizarro, O., & Montecinos, A. (2004). Interdecadal variability of the thermocline along the west coast of South America. *Geophysical Research Letters, 31*(20), 307–311.

Pizarro, O., Shaffer, G., Dewitte, B., & Ramos, M. (2002). Dynamics of seasonal and interannual variability of the Peru-Chile undercurrent. *Geophysical Research Letters, 29*(12), 22–21.

Pizarro-Koch, M., Pizarro, O., Dewitte, B., Montes, I., Ramos, M., Paulmier, A., & Garçon, V. (2019). Seasonal variability of the southern tip of the oxygen minimum zone in the eastern south pacific (30°–38°S): A modeling study. *Journal of Geophysical Research: Oceans, 124*, 8574–8604.

Plaza, F., Salas, R., & Yáñez, E. (2018). Identifying ecosystem patterns from time series of anchovy (*Engraulis ringens*) and sardine (*Sardinops sagax*) landings in northern Chile. *Journal of Statistical Computation and Simulation, 88*(10), 1863–1881.

Polanyi, K. (2001). *The great transformation: The political and economic origins of our time* (2nd ed., p. 360). Boston, MA: Beacon Press.

Polasky, S., Carpenter, S. R., Folke, C., & Keeler, B. (2011). Decision-making under great uncertainty: environmental management in an era of global change. *Trends in Ecology & Evolution, 26*(8), 398–404.

Pörtner, H. O., Roberts, D., Masson-Delmotte, V., Zhai, P., Tignor, M., Poloczanska, E., Mintenbeck, K., Alegria, A., Nicolai, M., Okem, A., Petzold, J., Rama, B., & Weyer, N. M. (Eds.). (2019). *IPCC special report on the ocean and cryosphere in a changing climate* (p. 755). Geneva: IPCC Intergovernmental Panel on Climate Change.

Ramajo, L., Fernández, C., Núñez, Y., Caballero, P., Lardies, M. A., & Poupin, M. J. (2019). Physiological responses of juvenile Chilean scallops (*Argopecten purpuratus*) to isolated and combined environmental drivers of coastal upwelling. *ICES Journal of Marine Science, 76*(6), 1836–1849.

Roberts, K. M. (1995). Neoliberalism and the transformation of populism in Latin America: The Peruvian case. *World Politics, 48*(1), 82–116.

Rykaczewski, R. R., Dunne, J. P., Sydeman, W. J., García-Reyes, M., Black, B. A., & Bograd, S. J. (2015). Poleward displacement of coastal upwelling-favorable winds in the ocean's eastern boundary currents through the 21st century. *Geophysical Research Letters, 42*(15), 6424–6431.

Ryther, J. H. (1969). Photosynthesis and fish production in the sea. *Science, 166*(3901), 72–76.

Salinger, M. J., Renwick, J. A., & Mullan, A. B. (2001). Interdecadal Pacific oscillation and south Pacific climate. *International Journal of Climatology: A Journal of the Royal Meteorological Society, 21*(14), 1705–1721.

Schindler, D. E., & Hilborn, R. (2015). Prediction, precaution, and policy under global change. *Science, 347*(6225), 953–954.

Schneider, W., Donoso, D., Garcés-Vargas, J., & Escribano, R. (2017). Water-column cooling and sea surface salinity increase in the upwelling region off central-south Chile driven by a poleward displacement of the South Pacific High. *Progress in Oceanography, 151*, 38–48.

Serra, R., Akester, M., Bouchón, M., & Gutierrez, M. (2012). Sustainability of the Humboldt current large marine ecosystem. In K. Sherman & G. McGovern (Eds.), *Frontline observations on climate change and sustainability of large marine ecosystems* (Vol. 17, pp. 112–134). New York: United Nations Development Programme.

Shaffer, G., Pizarro, O., Djurfeldt, L., Salinas, S., & Rutllant, J. (1997). Circulation and low-frequency variability near the Chilean coast: Remotely forced fluctuations during the 1991–92 El Nino. *Journal of Physical Oceanography, 27*(2), 217–235.

Silva, C., Leiva, F., & Lastra, J. (2019). Predicting the current and future suitable habitat distributions of the anchovy (*Engraulis ringens*) using the Maxent model in the coastal areas off central-northern Chile. *Fisheries Oceanography, 28*(2), 171–182.

Silva, C., Yáñez, E., Barbieri, M. A., Bernal, C., & Aranis, A. (2015). Forecasts of swordfish (*Xiphias gladius*) and common sardine (*Strangomera bentincki*) off Chile under the A2 IPCC climate change scenario. *Progress in Oceanography, 134*, 343–355.

Silva, N., Rojas, N., & Fedele, A. (2009). Water masses in the Humboldt Current System: Properties, distribution, and the nitrate deficit as a chemical water mass tracer for Equatorial Subsurface Water off Chile. *Deep Sea Research Part II: Topical Studies in Oceanography, 56*(16), 1004–1020.

Smith, R. L. (1968). Upwelling. *Oceanography and Marine Biology: An Annual Review, 6*, 11–46.

Sobarzo, M., Saldías, G. S., Tapia, F. J., Bravo, L., Moffat, C., & Largier, J. L. (2016). On subsurface cooling associated with the Biobio River Canyon (Chile). *Journal of Geophysical Research: Oceans, 121*(7), 4568–4584.

Stotz, W. (1997). Las áreas de manejo en la ley de pesca y acuicultura: Primeras experiencias y evaluación de la utilidad de esta herramienta para el recurso loco. *Estudios Oceanológicos, 16*, 67–86.

Stotz, W. (2000). When aquaculture restores and replaces an overfished stock: Is the conservation of the species assured? The case of the scallop *Argopecten purpuratus* in northern Chile. *Aquaculture International, 8*(2–3), 237–247.

Stotz, W. (2018). La experiencia de Chile en estudios de ecología de comunidades aplicados al aprovechamiento sostenible y conservación de la biodiversidad marina costera: El difícil camino hacia una armonía entre el ambiente, los pescadores y las regulaciones en la pesca. *Revista Comunicaciones Científicas y Tecnológicas, 4*(1), 275–285.

Stotz, W., Arias, N., & Miranda, F. (2019). Por qué muchas inversiones en la pesca artesanal no generan el desarrollo deseado? Deficiencias en la consideración de la dinámica de los recursos y la forma de vida de los pescadores artesanales en la política pública. In D. Henriquez & L. Moncayo (Eds.), *Construyendo Realidad: Estudios y propuestas para el desarrollo de la Región de Coquimbo* (pp. 297–320). Coquimbo, Chile: IPP UCN/Instituto de Políticas Públicas Universidad Católica del Norte - Sede Coquimbo.

Tarazona, J., Gutiérrez, D., Paredes, C., & Indacochea, A. (2003). Overview and challenges of marine biodiversity research in Peru. *Gayana, 67*(2), 206–223.

Thiel, M., Macaya, E. C., Acuña, E., Arntz, W. E., Bastias, H., Brokordt, K., Camus, P. A., Castilla, J. C., Castro, L. R., Cortés, M., Dumont, C. P., Escribano, R., Fernandez, M., Gajardo, J. A.,

Gaymer, C. F., Gomez, I., González, A. E., González, H. E., Haye, P. A., Illanes, J. E., Iriarte, J. L., Lancellotti, D. A., Luna-Jorquera, G., Luxoro, C., Manriquez, P. A., Marín, V., Muñoz, P., Navarrete, S. A., Perez, E., Poulin, E., Sellanes, J., Sepúlveda, H. H., Stotz, W., Tala, F., Thomas, A., Vargas, C. A., Vasquez, J. A., & Vega, A. (2007). The Humboldt current system of northern-central Chile: Oceanographic processes, ecological interactions and socio-economic feedback. *Oceanography and Marine Biology: An Annual Review, 45*, 195–344.

Ulloa, O., Canfield, D. E., DeLong, E. F., Letelier, R. M., & Stewart, F. J. (2012). Microbial oceanography of anoxic oxygen minimum zones. *Proceedings of the National Academy of Sciences, 109*(40), 15996–16003.

Vargas, C. A., Lagos, N. A., Lardies, M. A., Duarte, C., Manríquez, P. H., Aguilera, V. M., Broitman, B. R., Widdicombe, S., & Dupont, S. (2017). Species-specific responses to ocean acidification should account for local adaptation and adaptive plasticity. *Nature Ecology & Evolution, 1*(4), 1–7.

Vásquez, J. A., Vega, J. A., & Buschmann, A. H. (2006). Long term variability in the structure of kelp communities in northern Chile and the 1997–98 ENSO. In R. Anderson, K. Brodie, E. Onsoyen, & A. T. Critchley (Eds.), *Eighteenth international seaweed symposium* (pp. 279–293). Dordrecht: Springer.

Vergara, O., Dewitte, B., Montes Torres, I., Garçon, V., Ramos, M., Paulmier, A., & Pizarro, O. (2016). Seasonal variability of the oxygen minimum zone off Peru in a high-resolution regional coupled model. *Biogeosciences, 13*, 4389–4410.

Von Brand, E., Abarca, A., Merino, G. E., & Stotz, W. (2016). Scallop fishery and aquaculture in Chile: A history of developments and declines. In S. E. Shumway & G. J. Parsons (Eds.), *Scallops: Biology, ecology, aquaculture, and fisheries. developments in aquaculture and fisheries science* (Vol. 40, pp. 1047–1072). Amsterdam: Elsevier.

Wang, C., Deser, C., Yu, J. Y., DiNezio, P., & Clement, A. (2017). El Niño and southern oscillation (ENSO): A review. In P. W. Glynn, D. P. Manzello, & I. C. Enochs (Eds.), *Coral reefs of the eastern tropical Pacific* (pp. 85–106). Dordrecht, Holland: Springer.

Weyland, K. (2001). Clarifying a contested concept: Populism in the study of Latin American politics. *Comparative Politics, 31*(1), 1–22.

Weyland, K. (2004). Neoliberalism and democracy in Latin America: A mixed record. *Latin American Politics and Society, 46*(1), 135–157.

Wyrtki, K. (1962). The oxygen minima in relation to ocean circulation. *Deep Sea Research and Oceanographic Abstracts, 9*(1–2), 11–23.

Yáñez, E., Barbieri, M. A., Silva, C., Nieto, K., & Espındola, F. (2001). Climate variability and pelagic fisheries in northern Chile. *Progress in Oceanography, 49*(1–4), 581–596.

Yáñez, E., Lagos, N. A., Norambuena, R., Silva, C., Letelier, J., Muck, K. P., San Martin, G., Benitez, S., Broitman, B. R., Contreras, H., Duarte, C., Gelcich, S., Labra, F. A., Lardies, M. A., Manriquez, P. H., Quijon, P. A., Ramajo, L., Gonzalez, E., Molina, R., Gomez, A., Soto, L., Montecino, A., Barbieri, M. A., Plaza, F., Sanchez, F., Aranis, A., Bernal, C., & Bohm, G. (2017a). Impacts of climate change on marine fisheries and aquaculture in Chile. In *Climate Change Impacts on Fisheries and Aquaculture: A Global Analysis* (pp. 239–332). Hoboken, NJ: Wiley.

Yáñez, E., Plaza, F., Sánchez, F., Silva, C., Barbieri, M. A., & Bohm, G. (2017b). Modelling climate change impacts on anchovy and sardine landings in northern Chile using ANNs. *Latin American Journal of Aquatic Research, 45*(4), 675–689.

Yeh, S. W., Kug, J. S., Dewitte, B., Kwon, M. H., Kirtman, B. P., & Jin, F. F. (2009). El Niño in a changing climate. *Nature, 461*(7263), 511–514.

Yu, J. Y., Kao, H. Y., Lee, T., & Kim, S. T. (2011). Subsurface ocean temperature indices for Central-Pacific and Eastern-Pacific types of El Niño and La Niña events. *Theoretical and Applied Climatology, 103*(3–4), 337–344.

Zavala, R., Gutiérrez, D., Morales, R., Grünwaldt, A., Gonzales, N., Tam, J., Rodríguez, C., & Bucaram, S. (Eds.). (2019). *Avances del Perú en la adaptación al cambio climático del sector pesquero y del ecosistema marino-costero* (Vol. 679, p. 65). Lima, Perú: Inter-American Development Bank.

Part II
Marine Ecosystem Models in the Latin American Coasts

Modelling the Northern Humboldt Current Ecosystem: From Winds to Predators

Jorge Tam, Adolfo Chamorro, and Dante Espinoza-Morriberón

1 Introduction

The Northern Humboldt Current Ecosystem (NHCE) is one of the most productive systems in the world in terms of fish production. Trade winds along the Peruvian coast generate a persistent coastal upwelling of cold, nutrient rich waters, which produce high phytoplankton and zooplankton concentrations (Tarazona & Arntz, 2001; Chavez & Messié, 2009). This high productivity supports large fish biomasses (e.g., anchovy *Engraulis ringens*). However, the NHCE also presents one of the most intense Oxygen Minimum Zone (OMZ) (Graco et al., 2007) which limits the habitat of several species (Diaz & Rosenberg, 2008; Bertrand et al., 2011).

Due to its location near the Equator, the NHCE is strongly affected by the inter-annual variability of El Niño and the Southern Oscillation (ENSO) (Messié & Chávez, 2011). During El Nino (EN) events there is an increase of sea surface temperature (TSM; ~+3 °C), more oxygenated waters (Arntz et al., 2006), low nutrient content (Barber & Chavez, 1983; Graco et al., 2007; Graco et al., 2017), and low surface chlorophylls (Cl-a) (e.g. EN 1997–98; Thomas et al., 2001; Carr et al., 2002). This reduction in productivity is associated with a high mortality of species of commercial importance, such as anchovy (Ñiquen & Bouchón, 2004) and top predators due to lack of food (Tovar & Cabrera, 1985). In contrast, during La Niña (LN) events, negative SST anomalies prevail <–0.8 °C (sensu Trasmonte & Silva, 2008), with higher upwelling due to intensification of trade winds. Anchovy is favored by LN, extending its spatial distribution and increasing biomass (Bouchón & Peña, 2008). In addition to the interannual variability of ENSO, the NHCE is also affected by longer timescale variability such as interdecadal changes associated

J. Tam (✉) · A. Chamorro · D. Espinoza-Morriberón
Laboratorio de Modelado Oceanográfico, Ecosistémico y del Cambio Climático (LMOECC), Instituto del Mar del Perú (IMARPE), Callao, Peru
e-mail: jtam@imarpe.gob.pe

M. Ortiz, F. Jordán (eds.), *Marine Coastal Ecosystems Modelling and Conservation*, https://doi.org/10.1007/978-3-030-58211-1_3

with regime shifts (Chavez, Ryan, Lluch-Cota, & Ñiquen, 2003), long-term trends driven by climate change (Gutiérrez, Echevin, Tam, Takahashi, & Bertrand, 2014) and centennial variability (Salvatteci et al., 2018).

Despite the importance and intense monitoring of the NHCE, the complex and nonlinear interactions between biological communities, high environmental variability, and anthropogenic pressures require the development of simulation models of this ecosystem, in order to assess the effects in structure and function of biological communities under different natural and anthropogenic scenarios. Also, for the implementation of the Ecosystem approach to Fisheries (EAF) it is necessary to develop ecosystem models capable of performing fisheries management strategies evaluation (MSE) including multispecies interactions.

In order to build an ecosystem End-to-End (E2E) model of the NHCE, atmospheric, oceanic, biogeochemical, biological, and socioecological submodels have to be developed and then coupled in one-way and two-way modes. For the NHCE (Fig. 1), the development of these submodels has been the task of the Oceanographic, Ecosystem and Climate Change Modeling Laboratory (LMOECC) of the Peruvian Marine Research Institute (IMARPE) during the 2000s, in cooperation with different institutions (e.g., IRD from France, ZMT from Germany, and FRA from Japan).

1.1 Atmospheric Modeling of the NHCE

In the Peruvian upwelling system, as in other upwelling systems of the world ocean, the alongshore surface wind is the main driver of the coastal upwelling, bringing cold and nutrient rich waters to the surface layer where primary production occurs.

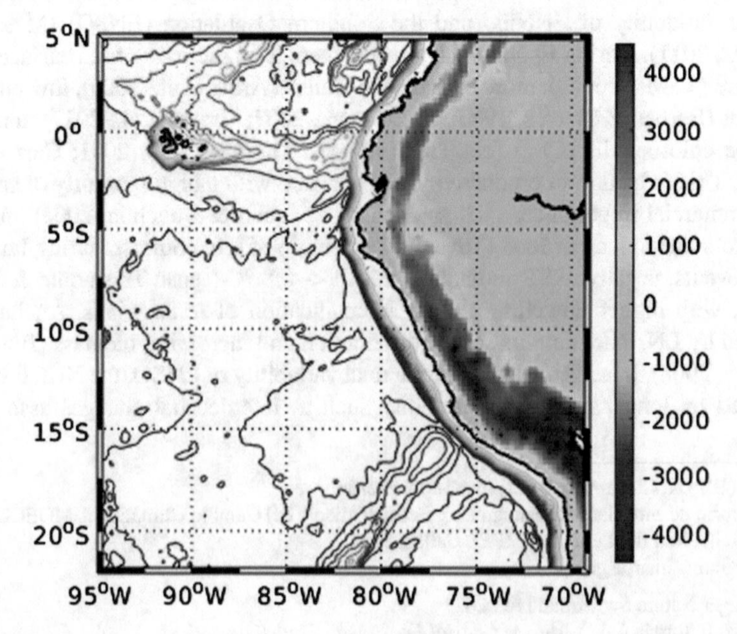

Fig. 1 NHCE domain

The Weather Research and Forecast (WRF) next-generation atmospheric model has been used to study the coastal wind variability and the processes responsible for it during the ocean surface layer warming conditions, at different time scales: (1) interannual time scales, corresponding to El Niño events; and (2) multi-decadal time scales resulting from regional climate change.

During typical El Niño events, an anomalous increase of the sea surface temperature (SST) occurs along the Ecuador and Peru coasts and the upwelling of cold water decreases (Fig. 2c; Colas, Capet, McWilliams, & Shchepetkin, 2008; Espinoza-Morriberon et al., 2017) despite an upwelling-favorable coast wind increase off the Peru coast is observed (Fig. 2a; Enfield, 1981; Kessler, 2006). This counter-intuitive wind increase during the 1997–1998 El Niño was studied by performing an atmospheric simulation forced by realistic oceanic (i.e., sea surface temperature) and lateral boundary conditions for the period 1994–2000 in order to analyze the physical processes involved in the coastal wind variability during an El Niño event. In addition, two model sensitivity experiments (BRY-EN and SST-EN) were performed in order to isolate the respective impacts of the local sea surface

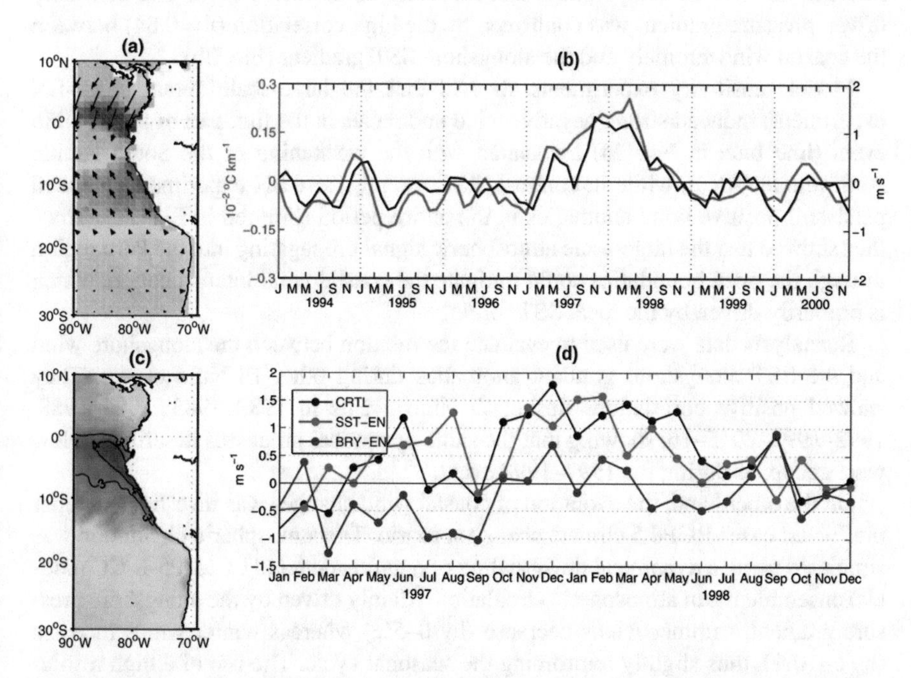

Fig. 2 (**a**) ERS wind anomalies (in m s^{-1}) off Peru and Northern Chile during El Niño conditions in November–February 1997/1998. Arrows mark the direction of the monthly wind anomalies. (**b**) Time evolution of alongshore SST gradient (in °C/25 km, positive equatorward, red line) and wind (in m s^{-1}, black line) anomalies. Anomalies were smoothed using a 3-month running mean, and averaged between 7 and 15°S and within 100 km from the coast. (**c**) Mean sea surface temperature (SST) anomalies (in °C) from OISST (over the same time period than in **a**). (**d**) Time series of coastal alongshore wind anomalies (in m s^{-1}) from CRTL (black line), SST-EN (red line), and BRY-EN (blue line) WRF model experiments (Chamorro et al., 2018)

temperature (SST) forcing and large-scale condition on the coastal wind intensification. The BRY-EN experiment was carried out using atmospheric boundary conditions from the years 1997–1998 (Niño boundary conditions) and SST forcing from the years 1994–1995 (so-called SST neutral conditions) to isolate the role of the large-scale signal. The SST-EN experiment was performed using atmospheric boundary conditions from the years 1994–1995 (neutral boundary conditions) and daily SST forcing from the years 1997–1998 (SST Niño conditions) to isolate the role of the SST local forcing.

A momentum balance analysis showed that the coastal wind intensification off the Peru coast was mainly driven by the enhancement of the alongshore pressure gradient. Vertical mixing tended to counterbalance the alongshore pressure gradient, leading to a quasi-equilibrium between the alongshore pressure gradient and the frictional force, consistently with previous modelling studies in the region (Belmadani, Echevin, Codron, Takahashi, & Junquas, 2014; Muñoz & Garreaud, 2005). The enhancement of the alongshore pressure gradient occurred because the atmospheric pressure decreased more north than south, in association with the larger increase of SST, air temperature, and humidity off northern Peru. The thermally driven pressure gradient was confirmed by the high correlation ($r = 0.84$) between the coastal wind anomaly and the alongshore SST gradient (Fig. 2b).

Model sensitivity experiments showed that the large-scale signal (BRY-EN experiment) induced strong negative wind anomalies in the first part of the El Niño event (line blue in Fig. 2d) associated with the weakening of the South Pacific Anticyclone (APS), while in contrast, SST forcing (SST-EN experiment) induced persistent positive wind anomalies in the entire period (line red in Fig. 2d). Thus, they showed that the large-scale atmospheric signal propagating into the Peru region through the open boundaries could mitigate the coastal wind intensification, which is primarily driven by the local SST forcing.

Reanalysis data were used to evaluate the relation between the alongshore wind and the SST alongshore gradient anomalies during other El Niño events. They showed positive correlations during El Niño events in 1982–1983, 1987–1988, 1992–1993, 2015–16, showing that the same dynamical processes described below were active, as during the 1997–1998 event.

On the other hand, the evolution of coastal wind changes was investigated under the "worst case" RCP8.5 climate change scenario. The atmospheric circulation was simulated using a dynamical downscaling of a multi-model (31 CMIP5-IPCC models) ensemble mean atmospheric circulation. Mainly driven by the alongshore pressure gradient, summer winds decrease (by 0–5%) whereas winter winds increase (by 5–10%), thus slightly reinforcing the seasonal cycle. The use of a high resolution nested grid (7 km) off Peru allows to simulate an enhanced local wind increase in winter in the Paracas Bay area. The wind changes are mainly associated with changes in the intensity and position of the South Pacific Anticyclone. The role of local factors such as land-sea surface temperature changes is shown to be negligible in our simulations (Fig. 3).

These results show that the increase of land-sea temperature contrast by global warming, a potential wind-driver local factor hypothesized by Bakun (1996), does not drive the wind changes off Peru coast, instead, they confirm that coastal winds could decrease off Peru in summer in agreement with previous regional climate change studies (Goubanova et al., 2011; Belmadani et al., 2014), although they projected weakened winds in winter. Thus regional climate change may induce a less productive summer season for the Peruvian region, due to less availability of nutrients (Echevin, Aumont, Ledesma, & Flores 2008).

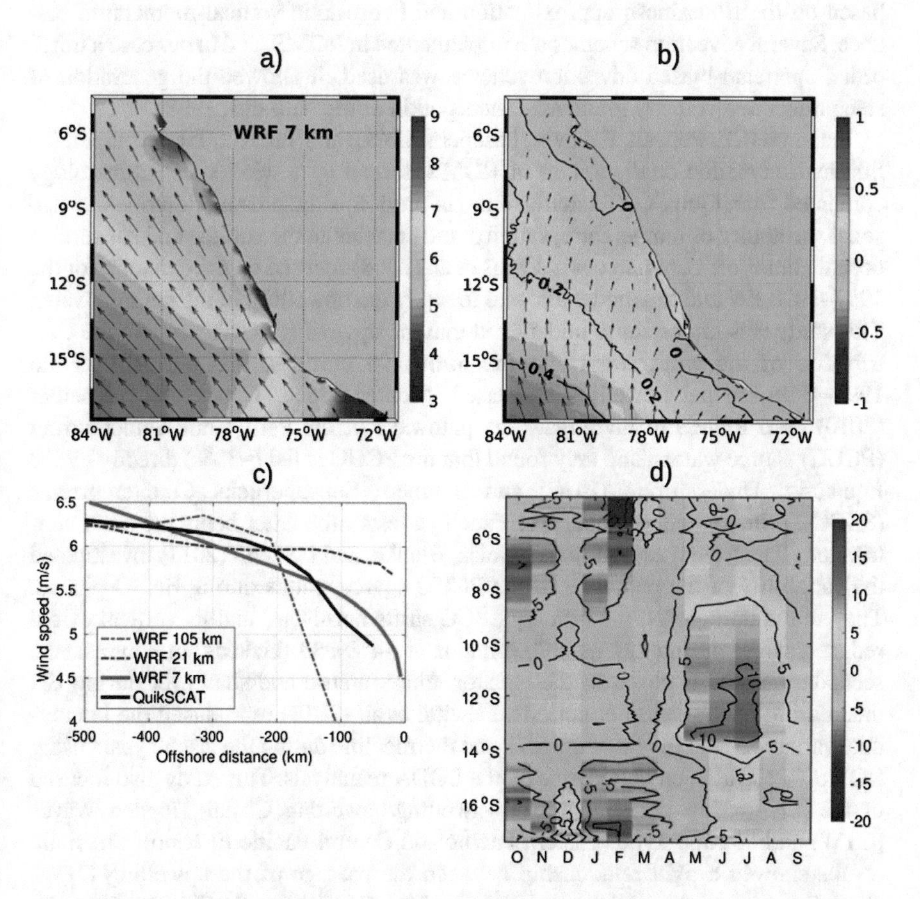

Fig. 3 (**a**) Mean surface wind for the period 2000–2003 from WRF7 simulation. (**b**) Change in mean surface wind (2086–2095 average minus 1994–2003 average) from WRF7. (**c**) Mean cross-shore surface wind speed between 7 and 13°S from WRF105, WRF21, WRF7, and QuikSCAT. (**d**) Relative change (in %) in the mean seasonal cycle of alongshore winds (in a coastal band of ~100 km) for the period 2085–2095 relative to the period 1994–2003 from WRF7

1.2 Oceanic Physical-Biogeochemical Modeling of the NHCE

The oceanic models allow to study processes in the ocean, which are often difficult to observe *in situ*. The possibility to formulate experiments to test different hypotheses makes them very useful tools. Both relatively simple and more complex physical-biogeochemical models have been used to investigate the main physical and biogeochemical processes and their interplay in the NHCE.

One of the physical model most widely used in the NHCE is the Regional Ocean Modelling System (ROMS) (Shchepetkin & McWilliams, 2005). ROMS simulates the ocean circulation and mixing. It is a split-explicit, free-surface oceanic model that resolves the Primitive Equations in an Earth-centered rotating environment, based on the Boussinesq approximation and hydrostatic vertical momentum balance. Several advection schemes are implemented in ROMS, and in our case a third-order, upstream-biased advection scheme was used. It allowed the generation of steep tracer and velocity gradients (Shchepetkin & McWilliams, 1998).

In the NHCE, Penven, Echevin, Pasapera, Colas, and Tam (2005) developed the first hydrodynamic configuration of ROMS, forced by a wind stress climatology computed from QuikSCAT satellite data, in order to simulate and describe the seasonal variability of temperature, salinity, and currents in the surface and subsurface oceanic fields off Peruvian coast. Colas et al. (2008) focused on the dynamics of the 1997–1998 EN and the model allowed to study the upwelling in the column water. This study was important to understand that the upwelling is triggered by the contribution of the wind stress and the horizontal currents, and during El Niño 1997–1998 the net upwelling decreased. Montes, Colas, Capet, and Schneider (2010) used ROMS to investigate the pathways of the Peru–Chile Undercurrent (PCUC) source waters, and they found that the PCUC is fed (~30%) directly by the Equatorial Undercurrent (EUC) and Southern Subsuperficial Countercurrents (SSCCs, primary and secondary, or Tsuchiya jests), the latter being the main contributors. Then, Montes, Wolfgang, Colas, Blanke, and Echevin (2011) investigated the variability of the pathways of the PCUC source waters during ENSO phases. They found during EN, the primary SSCC shifts northward and its vertical extent reduces, while during LN its core remains at ~4°S and thickens. In contrast, the secondary SSCC is closer to the equator, more intense and shallower during EN than during neutral and LN periods. Dewitte et al. (2012) reproduced the interannual variability and trends of the SST and thermocline during the last 50 years using ROMS forced at open boundaries by the SODA reanalysis. This study also focused in the relationship between the downwelling/upwelling Costal Trapped Waves (CTW) and El Niño Type (Eastern Pacific and Central Pacific El Niño). The main results showed a tight relationship between the passage of the upwelling CTWs along Peruvian coasts and the occurrence of the Central Pacific El Niño. Echevin, Colas, Chaigneau, and Penven (2011) used three different OGCMs to initialize and force the ROMS model in the NHCE, finding that the remote equatorial waves impact the amplitude and phase of seasonal mesoscale activity. Belmadani, Echevin, Dewitte, and Colas (2012) studied the propagation of the intraseasonal CTW and their impact on Rossby waves and mesoscale eddies, while Colas, Capet, McWilliams, and Li (2013) focused on the role and impact of the mesoscale variability on the cross-shore transport of heat.

The regional impacts of climate change on the physical oceanic conditions of the NHCE have been projected using the IPSL-CM4 global model downscaled with the ROMS model by Echevin, Goubanova, Belmadani, and Dewitte (2012). They found an enhanced stratification of the water column and an increase of coastal SST under a pessimistic 4xCO2 scenario. Later, Oerder et al. (2015) used ROMS model for dynamical downscaling of climate scenarios from the IPSL-CM4 global model, finding that while the coastal poleward undercurrent is intensified, the surface equatorial coastal jet shoals and the nearshore mesoscale activity are reinforced. Both studies found that reduction in alongshore wind stress and nearshore wind stress curl drive a year-round reduction in upwelling intensity.

Biogeochemical models of the NHCE began with carbon and nitrogen budgets. Walsh and Dugdale (1971) developed a simulation model of the nitrogen flowing through an upwelling plume, and Walsh (1981) compared the carbon budgets before and after anchovy overfishing in the periods 1966–69 and 1976–79, finding that the decline of anchovy led to increase in plankton, sardine, and hake, with increased carbon loading, sulfate reduction and decline in oxygen and nitrates. Carr (2003) used a size-based carbon 1D-vertical NPZD flow model to assess the response of plankton components to the forcing associated with normal, EN, and LN conditions, finding that during EN the plankton community decrease is driven by the new nitrate decrease of the upwelling source water, while during LN the phytoplankton increase is driven by the depth of the upper layer and upwelling rate. Jahncke, Checkley, and Hunt (2004) used an empirical model of carbon transfer through a simplified food web in the Peruvian upwelling system with and without the industrial fishery for anchovies, finding that the drastic decline in seabird abundance since the mid-1960s was likely due to competition for food with the fishery.

In last two decades, the international community of ROMS users has coupled the physical oceanic model to biogeochemical models of growing complexity. First studies coupled ROMS with simple Nitrate-Phytoplankton-Zooplankton-Detritus models (NPZD) (Kone et al., 2005), which simulates the low trophic level dynamics of the primary and secondary production. It takes into account simple relationships between phytoplankton and zooplankton and assumes that the phytoplankton growth is driven by nitrate availability and zooplankton predation. Nutrient loss is considered to be primarily from phytoplankton uptake and detritus comes from phyto- and zooplankton mortality (Gruber et al., 2006). More complex biogeochemical models such as Biogeochemical model for Eastern Boundary Upwelling Systems (BioEBUS, Gutknecht et al., 2013) and the Pelagic Interaction Scheme for Carbon and Ecosystem Studies (PISCES, Aumont, Ethé, Tagliabue, Bopp, & Gehlen, 2015) were also coupled to ROMS. BioEBUS is a nitrogen-based model, which also simulates the cycle of oxygen, phytoplankton, and zooplankton in regional areas. BioEBUS model includes a parameterization of processes as ammonification/nitrification and nitrification/denitrification/anammox, which occur under oxic and suboxic conditions, respectively. On the other hand, PISCES simulates the

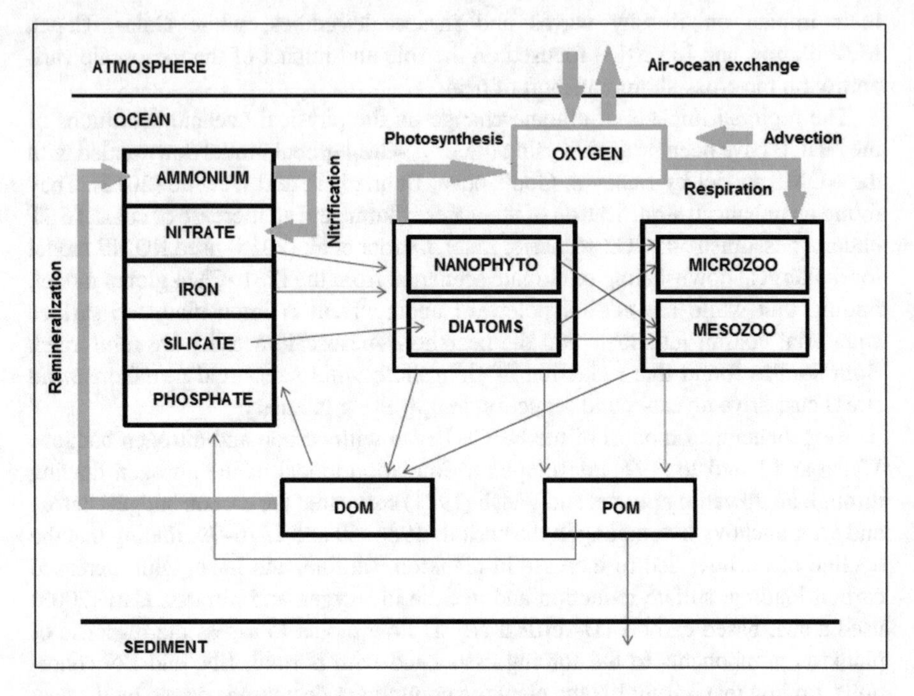

Fig. 4 Structure of the PISCES model. Stocks are marked by boxes and fluxes by arrows (Espinoza-Morriberon et al., 2016)

marine biological productivity and describes the biogeochemical cycles of carbon and the main nutrients (nitrate, phosphate, silicate, and iron), as well as the dissolved oxygen (Aumont et al., 2015; Resplandy et al., 2012; Fig. 4).

First studies in the NHCE using ROMS-PISCES were done by Echevin et al. (2008) who found that deeper MLD controls the phytoplankton decrease during winter. During spring and fall the productivity increase was produced by restratification and destratification of the water column, respectively. During winter, Fe was the limiting nutrient. Albert, Echevin, Lévy, and Aumont (2010) found that the absence of nearshore wind drop-off suppresses Ekman pumping and deepens the PCUC. A deeper PCUC impacts on the nutrient content (i.e., Fe) of the upwelled water, thus on the productivity.

The NHCE has been modeled also using the ROMS-BioEBUS coupled model, for example Montes et al. (2014) comparing different OGCMs (physical OBCs) found that the mean circulation presented small differences; however, the OMZ showed significant differences between them; and in the presence of an intense SSCC the OMZ shrinked due to an enhanced eastward transport of oxygenated waters. Using also the ROMS-BioEBUS model, Bettencourt et al. (2015) found that mesoscale processes delimit and maintain the OMZ boundaries and inject oxygen within the OMZ; Vergara et al. (2016) found that the seasonality of the coastal OMZ was related to seasonality of winds and productivity, the seasonal cycle of the offshore DO eddy flux and the coastal OMZ were in phase, while at the northern OMZ boundary, the DO eddy flux peaked in winter, and at southern boundary it peaked during spring; Mogollón and Calil (2017) found that there was an attenuation and expansion/intensification of the OMZ during EN and LN, respectively, during EN and LN the denitrification/anammox decrease and increase, respectively, on the other hand the removal nitrogen in the ocean was controlled by denitrification during EN and by anammox during LN; Yonss, Dietze, and Oschlies (2017) found that there was a decoupling between nitrite production and total nitrate change within eddies, the nutrient signature within the eddy was due to the presence of water masses from different origins.

The NHCE has been modeled also by Yang, Gruber, Long, and Vogt (2017) using the CEMS (Community Earth System Model, Gent et al., 2011) and BEC (Biological Elemental Cycling, Moore, Doney, & Lindsay, 2004), finding that the denitrification rate during LN (EN) was higher (lower) than during neutral periods, and the vari-

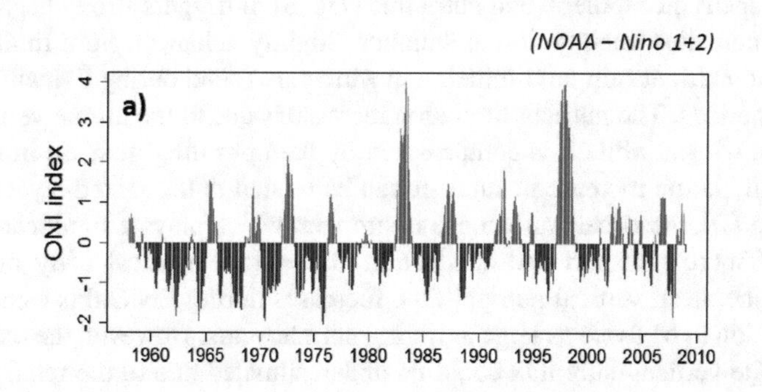

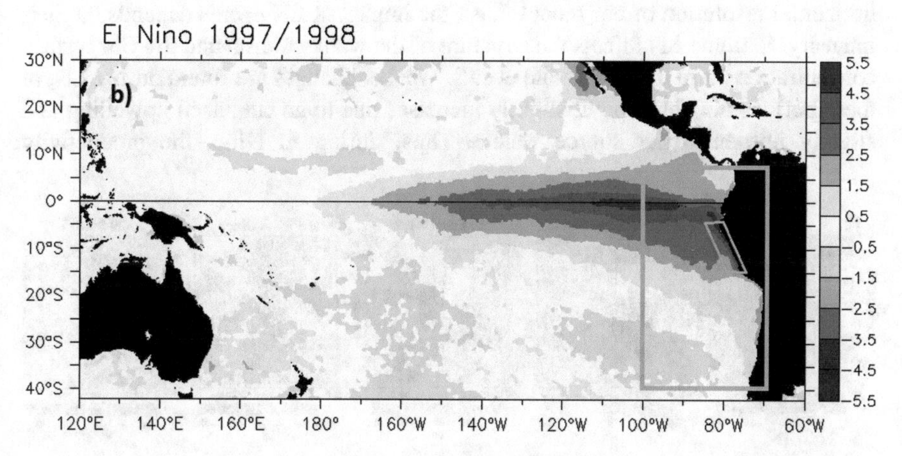

Fig. 5 ONI index computed in El Niño 1+2 region from 1958 to 2008 (**a**). This index was used to identify El Niño and La Niña events in Espinoza-Morriberon et al. (2017, 2019). Mean SST anomalies (Nov 1997–Jan 1998) of the Pacific Ocean from NOAA OISSTv2 (**b**). Gray box represents the simulated region in Espinoza-Morriberon et al. (2017, 2019) and the green box the analyzed area

ability of oxygen in the water column explained 90% of the variability of denitrification.

Recently, Espinoza-Morriberon et al. (2017, 2019) used ROMS-PISCES model to reproduce the main characteristics of different EN events from 1958 to 2008 (Fig. 5): the SST, Sea Surface Height (SSH), and oxygen increase, and the chlorophyll (Chl) and nutrient decrease off Peru. The wind-driven upwelling intensifies, but is partly compensated by an onshore geostrophic flow associated with an alongshore sea-level gradient. The compensating current is intense in winter and spring, and weaker or nonexistent in summer. The nitrate and iron contents of the upwelling source waters (SW) strongly decrease while their depths are little modified in the upwelling region. Besides, the SW properties in the equatorial region away from the coasts (88°W) do not change during neutral, EN moderate, and LN period, while during extreme EN events their nutrient content is lower (20%) probably because of the longer duration of the events (~16 months). The passage of strong downwelling coastal trapped waves (CTWs) increases the SSH and the SST, while it deepens the isotherm and nutricline (Fig. 6). It triggers strong negative surface Chl anomalies, mainly during summer. Slightly enhanced light limitation in summer and nutrient (nitrate) limitation in winter are found during EN with respect to neutral periods. The nutrient limitation increase is due to the nitrate vertical flux decrease in winter, while it is compensated by the upwelling increase in summer. The light limitation increase in summer can be related to the mixed-layer deepening. During EN, mesoscale turbulence is stronger, which plays a significant role in nutrient offshore transport and subduction. The nitrate vertical eddy flux, with respect to the mean vertical nutrient flux, increases during EN and is estimated in our simulation to be twice as large as during normal years. However, the magnitude of the nitrate vertical eddy flux could be underestimated due to the relatively low horizontal resolution of our model. Last, the impact of EN events depends on their intensity. Extreme EN affects the structure of the water column and the Chl surface concentration more than moderate events. Weaker changes are found during LN, in these periods the chlorophyll slightly increases due to an enhanced upwelling and slightly nutrient-richer source waters. Thus, during El Niño, the productivity

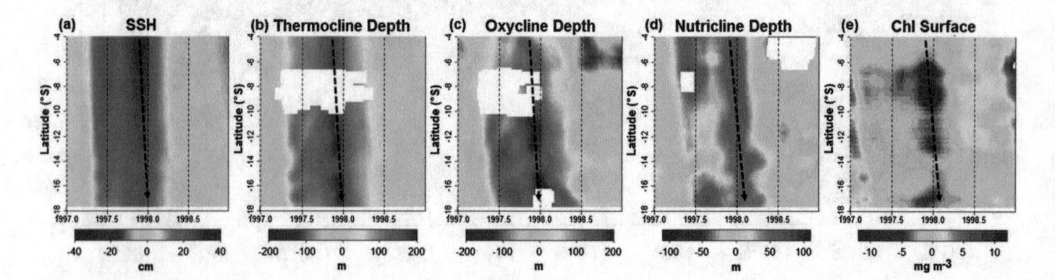

Fig. 6 Hovmöller latitude vs time of the anomalies of Sea Surface Height (**a**, in centimeters), thermocline depth (**b**, in meters and defined as 15 °C isotherm depth), oxycline depth (**c**, in meters, defined as isoline of 22 μmol kg^{-1} depth), nutricline depth (**d**, in meters, defined as isoline of 23 mL L^{-1} depth), and chlorophyll surface (**e**, in mg m^{-3}). The variables were averaged from 100 km to the coast and a 60-days filter was applied to them

decreases due to nutrient depletion associated with the deepening of the thermocline, the passage of intense downwelling Coastal Trapped Waves, and to an enhanced light limitation in summer.

Also using ROMS-PISCES model, Echevin et al. (2014) studied the propagation of intraseasonal CTW, finding that downwelling (upwelling) CTW deepens (shoals) the thermocline and nutricline. A deeper (shallower) nutricline triggers a productivity decrease (increase). The propagation of SSH is faster than the propagation of thermocline, nutricline, and surface chlorophyll. On the other hand, Mogollón and Calil (2018a) using ROMS-PISCES model found that an increase of the winds (~40%) led to a productivity increase (5%); this could be due to the decrease of the nutrient limitation relaxation (~10%), but nutrients were used less efficiently (decrease ~15%) due to enhanced eddy-driven offshore transport of nutrients. Mogollón and Calil (2018b) using ROMS-PISCES model found that dissolved inorganic carbon and SST drove the seasonal variability of the CO_2 fluxes at sea–air interface.

Espinoza-Morriberón et al. (2019) studied the physical and biogeochemical processes involved in the OMZ variability during EN and LN phases. The hydrodynamical-biogeochemical ROMS-PISCES coupled model was run for the period 1958–2008 in order to simulate the OMZ interannual variability. The model reproduced well the variability of the oxycline depth; however, within the OMZ the model underestimated the observed oxygen concentrations. The passage of downwelling and upwelling CTWs during EN and LN triggers a deepening (Fig. 6c) and shoaling of the oxycline, respectively. A high correlation was found between the depth of the isotherm and the oxycline in the nearshore region ($r \sim 0.8$), but the correlation drops offshore and south of 18°S, where the OMZ vanishes. The equatorial source waters (SW) are mainly transported eastward by the secondary Tsuchiya jet and the EUC during LN and EN, respectively, before reaching the Peruvian coast. During EN the SW are more oxygenated as they originate from the north of the OMZ and transit faster to the coastal region than during LN. Regarding the vertical eddy oxygen flux, it decreases during EN, while it does not change during LN. The horizontal eddy flux tends to inject oxygen into the OMZ at its boundaries (mainly at the southern one). During EN, this horizontal eddy flux strongly increases due to the eddy activity increase, intensifying eddy-driven horizontal ventilation. However, during LN, it is not modified. Remineralization is the main process consuming oxygen in the water column, followed by zooplankton respiration and nitrification. Nearshore, negative, and positive anomalies predominate during LN and EN, respectively, while ~30 km offshore and above 15 m, opposite sign patterns are found. Indeed, dissolved oxygen (DO) changes during LN and EN are influenced not only by physical processes but also by changes in DO production related to primary productivity changes above the upper limit of the OMZ, between 0 and ~20 m. The nearshore DO changes present an oxygen increase (decrease) consumption by remineralization and respiration during EN (LN). Nitrification is enhanced at depth (50–150 m) due to DO availability during EN. Sensitivity experiments performed to evaluate the influence of the equatorial remote (e.g., CTWs) and local forcings (e.g., winds) demonstrated that the equatorial remote forcing is the main driver of changes in the ZO2 and OMZ core during both EN and LN. During LN, the local wind fluctuations play a minor role, slightly increasing the ventilation of the surface layer and biological production of oxygen due to more upwelling and vertical mixing. During EN, the effect of the wind variability is slightly stronger: coastal upwelling is enhanced due to the alongshore wind increase associated with the SST warming, which partly compensates the ZO2 deepening driven by the downwelling CTWs, the surface layer becomes more ventilated as the oxycline deepens in association with the thermocline.

At a larger timescale, Espinoza (in prep.) used physical-biogeochemical models off Peru to analyze the interdecadal changes in oceanographic conditions from 1958 to 2008. Preliminary analyses showed that during the last decades, the large-scale remote forcing associated with equatorial variability mainly drives the summer chlorophyll increase and progressive deoxygenation trends, whereas local winds play a minor role.

The impacts of climate change on the biogeochemical conditions of the NHCE have been explored by Echevin et al. (2019) using ROMS-PISCES model under RCP8.5 scenario. They used 3 different Earth System Models (GFDL, IPSL, and CNRM) as boundary conditions. The biogeochemical downscaling of the three models indicated that under warming condition of the last decade of twenty-first century, the water column would present a deoxygenation and the subsurface water will present a decrease of chlorophyll concentration, although the surface water will be more productive. However, more realistic representations of the Earth System Models (ESM) are necessary to improve the projections of the future oceanographic conditions in the NHCE.

Future works on ocean physical-biogeochemical modeling will be devoted to simulate coastal areas, to disentangle the role local and remote forcings in submesoscale processes (Arellano et al. in prep.), which are important for small-scale fisheries, aquaculture and pollution control.

1.3 Individual Based Ecosystem Models of the NHCE

Lett, Penven, Ayón, and Fréon (2007) used the Ichthyop software, a Lagrangian Individual Based Model forced by water velocity fields from the three-dimensional hydrodynamical ROMS model to simulate the enrichment, concentration, and retention processes postulated by Bakun (1996) for the Peruvian anchovy *Engraulis ringens*. Brochier, Lett, Tam, Freon, and Colas (2008) investigated the factors driving the variability of eggs and larvae using the Ichthyop software, and they found that the most important factors were the buoyancy of the eggs and the larval vertical swimming behavior, which reproduced the observed spatio-temporal patterns, with higher retention between 10°S and 20°S, and higher eggs and larvae surface retention in winter, but higher subsurface retention in summer. Brochier et al. (2009) used an evolutionary Ichthyop model in order to investigate the reproductive strategies of small pelagics in three upwelling systems (NHCE, Benguela, and Canarias), and they found that in the NHCE (6°S–14°S), the retention over the shelf and the avoidance of dispersive structures were the factors which best reproduced the observed spatio-temporal spawning patterns. In addition, the shelf retention constraint led to selection of a particular spawning season during the period of minimum upwelling in all three of the upwelling regions.

The impact of climate change on early stages of small pelagic fishes was explored by Brochier et al. (2013) who used Ichthyop model to simulate larval retention at nursery areas under pre-industrial, $2xCO_2$ and $4xCO_2$ downscaled scenarios from the IPSL-CM4 global model (Echevin et al., 2012). They found an increase in larval retention due to higher stratification of the water column caused by the regional warming, but they also found a decrease of the nursery area due to lower plankton productivity and shoaling of the oxycline, the combination of these processes would result in a net reduction of fish recruitment. A different IBM anchovy model was developed by Xu, Chai, Rose, Niquen, and Chavez (2013) to study the influence of physical and biological processes on Peruvian anchovy recruitment. ROMS was coupled to CoSiNE (Carbon Silicate Nitrogen Ecosystem) model to force the IBM, the model reproduced the observed spatial changes in anchovy distribution during El Niño, simulating survivors were located farther south, closer to shore, and deeper than in normal years.

The Eulerian Spatial Ecosystem And Populations Dynamics Model (SEAPODYM-SP) was modified by Hernandez et al. (2014) to simulate the spawning habitat and eggs and larvae of anchovy and sardine. The model was forced by ROMS-PISCES outputs, and reproduced the observed distribution of anchovy larvae further south than the distribution of sardine larvae in the northern Peruvian shelf. Marzloff, Shin, Tam, Travers, and Bertrand (2009) used the OSMOSE (Object-oriented Simulator of Marine ecOSystEms) model, an ecosystem individual based model, to simulate the dynamics and life cycle of eight major species of the NHCE. OSMOSE is a species- and age-based model with size-based predation. They explored different fishery scenarios of Peruvian hake (*Merluccius gayi peruanus*). Later, Oliveros-Ramos, Verley, Echevin, and Shin (2017) tested different calibration schemes to estimate parameters of OSMOSE coupled to ROMS-PISCES using an evolutionary algorithm to improve the model fit to time series data of NHCE.

A future development on IBMs is to link Ichthyop model to processes at different scales: physical and biogeochemical regional dynamics, larval dispersion, fish dynamic energy budget (DEB), and swimming behavior (Brochier et al., 2018). A DEB model for Peruvian scallop has been developed by Aguirre-Velarde et al. (2019) and laboratory experiments for parameterizing a Peruvian anchovy DEB model are underway (A. Aguirre pers. comm.).

1.4 Ecotrophic Modelling of the NHCE

The first mass balance models were developed by Jarre-Teichmann, Muck, and Pauly (1989), Jarre and Pauly (1993) to compare food web structure during the decades of 1953–1959, 1960–1969, 1973–1979. Guenette et al. (2008) used the 1950 model of Jarre et al. (1991) and fitted the model with time series of biomasses and catches available for the period 1953–1984; however, they recommended that models should accommodate all the time series available for a given system (the old, contrast-rich series, and the more precisely estimated new ones). Tam et al. (2008) compared the effects of 1995–96 La Niña (LN) and 1997–98 El Niño (EN) events by building steady-state food web models using the Ecopath with Ecosim (EwE) software (Christensen & Pauly, 1992). For these models the area covered 4°S–16°S and up to 60 nm off the coast. The model included 32 functional groups, including species that became important in recent years (e.g., jumbo squid and mesopelagics). Analyzing ecosystem networks indicators, EN impacts on food web structure showed a shrinking of ecosystem size in terms of energy flows (Fig. 7), use of alternate pathways leading to more zooplankton dominated diets and higher trophic levels, which stressed the need for precautionary management of fisheries during and after EN.

Taylor et al. (2008) used the 1995–96 La Niña steady-state model of Tam et al. (2008), and time series from 1995 to 2004 of biomass, catch, fisheries mortality, and fishing effort from IMARPE to assess the contributions of several external drivers (i.e., phytoplankton changes, immigration of mesopelagic fishes, and fishing rate)

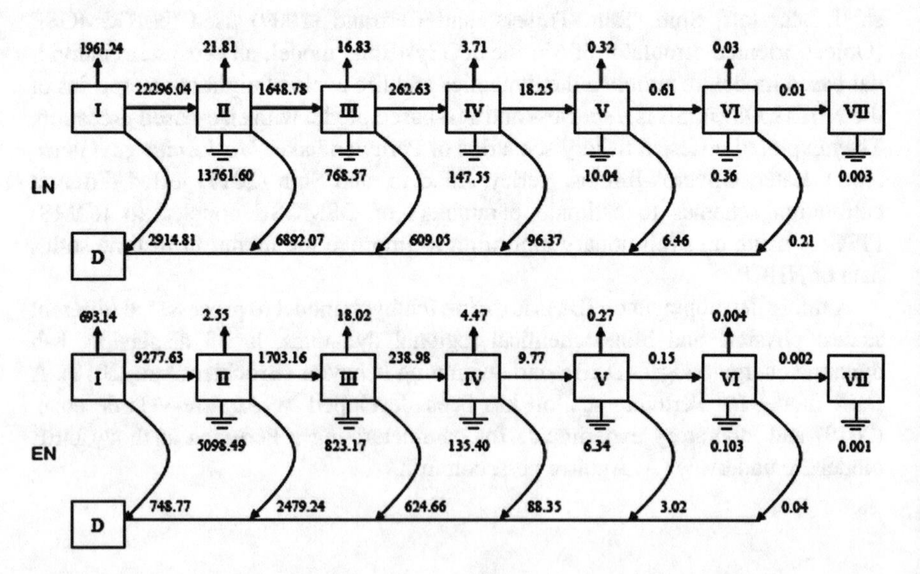

Fig. 7 Comparison of trophic structures in LN and EN: NHCE canonical trophic food chains. Flow networks were aggregated into equivalent trophic chains with distinct trophic levels. Flows are in t km^2 y^1, straight arrows indicate exports, ground symbols indicate respirational losses, and curved arrows indicate returns to detritus

over an ENSO cycle, using the EwE software (Christensen & Walters, 2004). Fishing rate was the most important driver, followed by immigration of mesopelagic fishes up to five years after EN, and finally phytoplankton was important during EN and the subsequent three years after EN. Internal control settings showed bottom-up forcing of meso- and macro-zooplankton to their predators anchovy and sardine, as well as bottom-up forcing of anchovy and sardine to their predators seabirds and pinnipeds; however, a "wasp-waist" control by small pelagic fish was not supported.

Tam et al. (2008) and Taylor et al. (2008) models formed the basis for other ecosystem studies focusing on different target groups such as hake (Tam et al., 2009), anchovy (Tam et al., 2010), and jellyfishes (Chiaverano et al., 2018); the ecosystem model was also coupled to a supply chain model in order to perform a sustainability assessment of different exploitation scenarios (Avadí, Fréon, & Tam, 2014). Tam et al. (2009) simulated fishing scenarios for hake in order to explore different recovery biomass projections. Later, Tam et al. (2010) found fishing mortality levels of anchovy associated with minimum spawning biomass taking into account its multispecific relationships. Chiaverano et al. (2018) found that a simulated increase in forage fish harvest enhanced jellyfish productivity, thus supporting the hypothesis that forage fishing provides a competitive release for large jellyfish in the NHCE; accordingly a simulated jellyfish bloom resulted in a decline in productivity of all functional groups, including forage fish.

Avadí et al. (2014) coupled the ecotrophic model of Tam et al. (2008) using EwE with the material and energy-flow network model using Umberto, to trace various local and global supply chains derived from the Peruvian Anchovy fishery, especially reduction of Anchovy into fishmeal and oil. Sustainability indicators (e.g., environmental, socio-economic, and nutritional) were measured to compare the performance of supply chains modeled under different policy-based scenarios. Three scenarios were explored: (i) status quo of fish exploitation, (ii) increase in Anchovy landings for direct human consumption and (iii) radical decrease in total Anchovy landings to allow other fish stocks to prosper. They found that the second scenario provided the best performance in sustainability indicators of the ecosystem—supply chain systems.

The NHCE food web models have also been compared with other ecosystems, particularly among other eastern boundary current systems, giving useful insights about multispecies management. Jarre-Teichmann et al. (1989) models were compared with other upwelling ecosystems. Jarre-Teichmann and Christensen (1998) found common global properties in upwelling ecosystems: (i) general species composition and major flow patterns, (ii) important role of small pelagic fish, (iii) the total fraction of primary production required to sustain the fish groups in the ecosystem, (iv) general correlation between total catch and the trophic level of the fishery, and the primary production, (v) low mean transfer of energy up the food web, and (vi) overall low system maturity. Jarre-Teichmann, Shannon, Moloney, and Wickens (1998) found that the energy transfer from primary producers to small pelagic fish was less efficient in the southern Benguela than in Peru, suggesting that as opportunistic generalists, the species in the southern Benguela are adapted to variability, and as such they are survivors, not optimal exploiters of their upwelling ecosystem. Jarre-Teichmann (1998) found that fishery took 20–30% of the production of dominant fish species in Peru, Benguela, and Canarias, but less than 10% in California, where a larger fraction was consumed by top predators because they form the basis of the tourism industry. Moloney et al. (2005) concluded that between-system differences in indicators may be more important than differences that could be attributed to heavy vs. medium fishing impacts. Tam et al. (2008) models were compared with other ecosystems (i.e., California current, northern Humboldt, North Sea, southern Benguela, and southeast Australia) to explore the multispecies effects of fishing low trophic species (Smith et al., 2011). Also important will be the comparative modeling between NHCE and Kuroshio Current ecosystem to understand synchronous alternations of small pelagic fish species (Oozeki et al., 2019).

2 Conclusions

Physical and biogeochemical models have been used to understand the oceanographic dynamics of the NHCE. The first studies focused to describe and understand the seasonal and interannual variability of the SST and currents in the water column, mainly during El Niño and La Niña events, while during the last years the development of computational capability allowed to investigate mesoscale and submesoscale processes in detail (e.g., eddies and front dynamics). ROMS model was coupled to biogeochemical models with different level of complexity, such as NPZD, BioEBUS, and PISCES. These coupled models were used to understand the seasonal and interannual variability of the Chl and the oxygen concentration in the water column. In recent years, first efforts to investigate the sensibility of NHCE to climate change scenarios stressed the importance that ESM needs to reproduce better the current conditions of the physical and biogeochemical ocean dynamics.

The use of simulation models in the NHCE allowed to test new hypotheses and overcome old paradigms about the structure and functioning of the ecosystem. Without any doubt, the use of E2E models will be increasingly necessary both for projections of impacts of climate change as well as for inclusion of the human dimension (i.e., socioeconomics) and evolutionary dynamics into these models in order to achieve the ecosystem approach to fisheries and the sustainable development of socioecological systems under natural and anthropogenic stressors.

Acknowledgements We would like to thank our partners during the implementation of several models for the NHCE in the LMOECC: Pierrick Penven, Vincent Echevin, and Francois Colas during the implementation of WRF and ROMS-PISCES models; Marc Taylor, Matthias Wolff, and Moritz Stäbler during the implementation of EwE; Timothée Brochier and Jorge Flores during the implementation of Ichthyop model; Martin Marzloff and Ricardo Oliveros during the implementation of OSMOSE model; Arturo Aguirre, Laure Pecquerie, and Takeshi Okunishi for the on-going support in the implementation of bioenergetic models. We also acknowledge the support of IRD and IADB for donating High Performance Computing Clusters to IMARPE, which allowed to run several ecosystem submodels for the NHCE. Finally, special thanks to the LMOECC staff (Cinthia

Arellano, Carlos Quispe, Yván Romero, and Jorge Ramos) as well as the observational staff from IMARPE which was crucial to validate and interpret the model outputs.

References

Albert, A., Echevin, V., Lévy, M., & Aumont, O. (2010). Impact of nearshore wind stress curl on coastal circulation and primary productivity in the Peru upwelling system. *Journal of Geophysical Research, 115*, C12033. https://doi.org/10.1029/2010JC006569.

Arntz, W. E., Gallardo, V. A., Gutíerrez, D., Isla, E., Levin, L. A., Mendo, J., Neir, C., Rowe, G. T., Tarazona, J., & Wolff, M. (2006). El Nino and similar perturbation effects on the benthos of the Humboldt, California, and Benguela Current upwelling ecosystems. *Advances in Geosciences, 6*, 243–265.

Aumont, O., Ethé, C., Tagliabue, A., Bopp, L., & Gehlen, M. (2015). PISCES-v2: An ocean biogeochemical model for carbon and ecosystem studies. *Geoscientific Model Development, 8*, 2465–2513. https://doi.org/10.5194/gmd-8-2465-2015.

Aguirre-Velarde, A., Pecquerie, L., Jean, F., Thouzeau,G., & Flye-Sainte-Marie, J. (2019). Predicting the energy budget of the scallop Argopecten purpuratus in an oxygen–limiting environment. *Journal of Sea Research, 143*, 254–261. https://doi.org/10.1016/j.seares.2018.09.011ff.ffhal02114544f.

Avadí, A., Fréon, P., & Tam, J. (2014). Coupled ecosystem/supply chain modelling of fish products from sea to shelf: The Peruvian Anchovy Case. *PLoS ONE, 9*, 1–21.

Bakun, A. (1996). *Patterns in the ocean: Ocean processes and marine population dynamics*. 7. https://doi.org/10.1016/s0278-4343(97)00037-x

Barber, R. T., & Chavez, F. P. (1983). Biological consequences of El Niño, *Sciences, 222*, 1203–1210. https://doi.org/10.1126/science.222.4629.1203.

Belmadani, A., Echevin, V., Codron, F., Takahashi, K., & Junquas, C. (2014). What dynamics drive future wind scenarios for coastal upwelling off Peru and Chile? *Climate Dynamics, 43*, 1893–1914. https://doi.org/10.1007/s00382-013-2015-2.

Belmadani, A., Echevin, V., Dewitte, B., & Colas, F. (2012). Equatorially forced intraseasonal propagations along the Peru–Chile coast and their relation with the nearshore eddy activity in 1992–2000: a modeling study. *Journal of Geophysical Research, 117*, C04025. https://doi.org/10.1029/2011JC007848.

Bertrand, A., Chaigneau, A., Peraltilla, S., Ledesma, J., Graco, M., Monetti, F., & Chavez, F. (2011). Oxygen: A fundamental property regulating pelagic ecosystem structure in the coastal southeastern tropical pacific, *PLOS ONE, 6*(12), e29558, https://doi.org/10.1371/journal.pone.0029558.

Bettencourt, J. H., López, C., Hernández-García, E., Montes, I., Sudre, J., Dewitte, B., Paulmier, A., & Garçon, V. (2015). Boundaries of the Peruvian Oxygen Minimum Zone shaped by coherent mesoscale dynamics, Nat. Geosci., 8, 937–940. https://doi.org/10.1038/ngeo2570.

Bouchón, M., & Peña, C., (2008). Impactos de los eventos La Niña en la pesquería peruana. Inf. Inst. Mar Perú. 35 (3): 193–198.

Brochier, T., Auger, P.-A., Pecquerie, L., Machu, E., Capet, X., Thiaw, M., et al. (2018). Complex small pelagic fish population patterns arising from individual behavioral responses to their environment. *Progress in Oceanography, 164*, 12–27.

Brochier, T., Colas, F., Lett, C., Echevin, V., Cubillos, L. A., Tam, J., et al. (2009). Small pelagic fish reproductive strategies in upwelling systems: A natal homing evolutionary model to study environmental constraints. *Progress in Oceanography, 83*, 261–269.

Brochier, T., Echevin, V., Tam, J., Chaigneau, A., Goubanova, K., & Bertrand, A. (2013). Climate change scenarios experiments predict a future reduction in small pelagic fish recruitment in the Humboldt Current system. *Global Change Biology, 19*, 1841–1853.

Brochier, T., Lett, C., Tam, J., Freon, P., & Colas, F. (2008). An individual-based model study of anchovy early life history in the northern Humboldt Current system. *Progress in Oceanography, 79*, 313–325.

Carr, M. -E., Strub, P. T., Thomas, A. C., & Blanco J. L. (2002). Evolution of 1996–1999 La Niña and El Niño conditions off the western coast of South America: A remote sensing perspective, *Journal of Geophysical Research, 107*(C12), 3236. https://doi.org/10.1029/2001JC001183.

Carr, M. E. (2003). Simulation of carbon pathways in the planktonic ecosystem off Peru during the 1997–1998 El Niño and La Niña. *Journal of Geophysical Research, 108*(C12), 3380. https://doi.org/10.1029/1999JC000064.

Chamorro, A. (2018). Coastal winds dynamics in the Peruvian upwelling system under warming conditions: impact of El Niño and regional climate change. Sorbonne Université. https://tel.archives-ouvertes.fr/tel-02163945.

Chamorro, A., Echevin, V., Colas, F., Oerder, V., Tam, J., & Quispe-Ccalluari, C. (2018). Mechanisms of the intensification of the upwelling-favorable winds during El Niño 1997-1998 in the Peruvian upwelling system. *Climate Dynamics*, 1–17. https://doi.org/10.1007/s00382-018-4106-6.

Chavez, F. P., Ryan, J. P., Lluch-Cota, S., & Ñiquen, C. M. (2003). From anchovies to sardines and back-Multidecadal change in the Pacific Ocean. *Science, 299*, 217–221.

Chavez, F., & Messié, M. (2009). A comparison of Eastern Boundary Upwelling Ecosystems. *Progress in Oceanography, 83*, 80–96. https://doi.org/10.1016/j.pocean.2009.07.032.

Chiaverano, L. M., Robinson, K. L., Tam, J., Ruzicka, J. J., Quiñones, J., Aleksa, K. T., et al. (2018). Evaluating the role of large jellyfish and forage fishes as energy pathways, and their interplay with fisheries, in the Northern Humboldt. *Current System Progress in Oceanography, 164*, 28–36.

Christensen, V., & Pauly, D. (1992). ECOPATH II-A system for balancing steady state ecosystem models and calculating network characteristics. *Ecological Modelling, 61*(3–4), 169–185.

Christensen, V., & Walters, C. (2004). Ecopath with Ecosim: Methods, capabilities and limitations. *Ecological Modelling, 172*(2–4), 109–139.

Colas, F., Capet, X., McWilliams, J. C., & Li, Z. (2013). Mesoscale eddy buoyancy flux and eddy-induced circulation in eastern boundary currents. *Journal of Physical Oceanography, 43*, 1073–1095. https://doi.org/10.1175/JPO-D-11-0241.

Colas, F., Capet, X., McWilliams, J. C., & Shchepetkin, A. (2008). 1997–98 El Niño off Peru: A numerical study. *Progress in Oceanography, 79*, 138–155.

Dewitte, B., Vazquez-Cuervo, J., Goubanova, K., Illig, S., Takahashi, K., Cambon, G., et al. (2012). Change in El Niño flavours over 1958–2008: Implications for the long-term trend of the upwelling off Peru. *Deep-Sea Research Part II, 77–80*, 143–156. https://doi.org/10.1016/j.dsr2.2012.04.011.

Diaz, R. J., & Rosenberg, R. (2008). Spreading dead zones and consequences for marine ecosystems. *Science, 321*, 926–929. https://doi.org/10.1126/science.1156401.

Echevin, V., Albert, A., Lévy, M., Aumont, O., Graco, M., & Garric, G. (2014). Remotely-forced intraseasonal variability of the Northern Humboldt Current System surface chlorophyll using a coupled physical-ecosystem model. *Continental Shelf Research, 73*, 14–30. https://doi.org/10.1016/j.csr.2013.11.015.

Echevin, V., Aumont, O., Ledesma, J., & Flores, G. (2008). The seasonal cycle of surface chlorophyll in the Peruvian upwelling system: A model study. *Progress in Oceanography, 79*, 167–176.

Echevin, V., Colas, F., Chaigneau, A., & Penven, P. (2011). Sensitivity of the Northern Humboldt Current System nearshore modeled circulation to initial and boundary conditions. *Journal of Geophysical Research, 116*, C07002. https://doi.org/10.1029/2010JC006684.

Echevin, V., Gévaudan, M., Colas, F., Espinoza-Morriberon, D., Tam, J., Aumont, O., et al. (2019). Physical and biogeochemical impacts of RCP8.5 scenario in the Peru upwelling system. *Biogeosciences Discussions*. https://doi.org/10.5194/bg-2020-4.

Echevin, V., Goubanova, K., Belmadani, A., & Dewitte, B. (2012). Sensitivity of the Humboldt Current system to global warming: A downscaling experiment of the IPSL-CM4 model. *Climate Dynamics, 38*, 761–774.

Enfield, D. B. (1981). Thermally driven wind variability in the planetary boundary layer above Lima, Peru. *Journal of Geophysical Research, 86*(C3), 2005–2016. https://doi.org/10.1029/JC086iCO3p02005.

Espinoza-Morriberon, D., Echevin, V., Colas, F., Tam, J., Ledesma, J., Graco, M., et al. (2017). Impact of the El Nino event on the productivity of the Peruvian Coastal Upwelling System. *Journal of Geophysical Research: Oceans, 122*(7), 5423–5444. https://doi.org/10.1002/2016JC012439.

Espinoza-Morriberon, D., Echevin, V., Romero, C. Y., Ledesma, J., Oliveros-Ramos, R., & Tam, J. (2016). Biogeochemical validation of an interannual simulation of the ROMS-PISCES coupled model in the Southeast Pacific. *Revista Peruana de Biología, 23*(2), 159–168. https://doi.org/10.15381/rpb.v23i2.12427.

Espinoza-Morriberón, D., Echevin, E., Colas, F., Tam, J., Gutierrez, D., Graco, M., et al. (2019). Oxygen variability during ENSO in the Tropical South Eastern Pacific. Frontiers in Marine Science. https://doi.org/10.3389/fmars.2018.00526.

Gent, P. R., Danabasoglu, G., Donner, L. J., Holland, M. M., Hunke, E. C., Jayne, S. R., et al. (2011). The Community Climate System Model version 4. *Journal of Climate, 24*, 4973–4991. https://doi.org/10.1175/2011JCLI4083.1.

Goubanova, K., Echevin, V., Dewitte, B., Codron, F., Takahashi, K., Terray, P., &Vrac, M. (2011). Statistical downscaling of sea-surface wind over the Peru–Chile upwelling region: diagnosing the impact of climate change from the IPSL-CM4 model. *Climate Dynamics, 36*(7–8), 1365–1378. https://doi.org/10.1007/s00382-010-0824-0

Graco, M., Ledesma, J., Flores, G.. & Girón, M. (2007). Nutrientes, oxígeno y procesos biogeoquímicos en el sistema de surgencias de la corriente de Humboldt frente a Perú. *Revista Peruana de Biología, 14*(1), 117–128.

Graco, M., Purca, S., Dewitte, B., Morón, O., Ledesma, J., Flores, G., Castro, C., & Gutiérrez, D. (2017). The OMZ and nutrient features as a signature of interannual and low-frequency variability in the Peruvian upwelling system, *Biogeosciences, 14*, 4601–4617, https://doi.org/10.5194/bg-14-4601-2017.

Gruber, N., Frenzel, H., Doney, S. C., Marchesiello, P., McWilliams, J. C., Moisan, J. R., et al. (2006). Eddy resolving simulation of plankton ecosystem dynamics in the California Current System. *Deep Sea Research, Part I, 53*(9), 1483–1516.

Gutiérrez, D., Echevin, V., Tam, J., Takahashi, K., & Bertrand, A. (2014). Impacto del cambio climático sobre el mar peruano: tendencias actuales y futuras. In A. Grégoire (Ed.), *El Perú frente al cambio climático. Resultado de investigaciones franco-peruanas* (pp. 142–155). IRD.

Gutknecht, E., Dadou, I., Le Vu, B., Cambon, G., Sudre, J., Garçon, V., et al. (2013). Coupled physical/biogeochemical modeling including O2-dependent processes in the Eastern Boundary Upwelling Systems: Application in the Benguela. *Biogeosciences, 10*, 3559–3591. https://doi.org/10.5194/bg-10-3559-2013.

Guenette, S., Christensen, V., Pauly, D. (2008). Trophic modelling of the Peruvian upwelling ecosystem: towards reconciliation of múltiple datasets. *Progress in Oceanography 79*, 352–365.

Hernandez, O., Lehodey, P., Senina, I., Echevin, V., Ayón, P., Bertrand, A., et al. (2014). Understanding mechanisms that control fish spawning and larval recruitment: Parameter optimization of an Eulerian model (SEAPODYM-SP) with Peruvian anchovy and sardine eggs and larvae data. *Progress in Oceanography, 123*, 105–122.

Jahncke, J., Checkley, D. M., & Hunt, G. L. (2004). Trends in carbon flux to seabirds in the Peruvian upwelling system: Effects of wind and fisheries on population regulation. *Fisheries Oceanography, 13*, 208–223.

Jarre, A., Muck, P., & Pauly, D. (1991). Two approaches for modelling fish stock interactions in the Peruvian upwelling ecosystem. In: N. Daanand, & M. P. Sissenwine, (Eds.), Multispecies Models Relevant to Management of Living Resources. ICES Marine Science Symposium 193, (pp. 171–184).

Jarre, A., & Pauly, D. (1993). Seasonal changes in the Peruvian upwelling ecosystem. In: V. Christensen, & D. Pauly, (Eds.), Trophic Models of Aquatic Ecosystems, ICLARM Conference Proceedings, vol. 26, (pp. 307–314).

Jarre-Teichmann, A. (1998). The potential role of mass balance models for the management of upwelling ecosystems. *Ecological Applications, 8*, S93–S103.

Jarre-Teichmann, A., & Christensen, V. (1998). Comparative modelling of trophic flows in four large upwelling ecosystems: Global versus local effects. In M.-H. Durand, P. Cury, R. Mendelssohn, C. Roy, A. Bakun, & D. Pauly (Eds.), *Global versus local changes in upwelling systems* (pp. 423–443). Paris: ORSTOM.

Jarre-Teichmann, A., Muck, P., & Pauly, D. (1989). *Interactions between fish stocks in the Peruvian upwelling ecosystem.* ICES 1989 MSM Symposium Paper 27, 24 p.

Jarre-Teichmann, A., Shannon, L. J., Moloney, C. L., & Wickens, P. A. (1998). Comparing trophic flows in the Southern Benguela to those in other upwelling ecosystems. *South African Journal of Marine Science, 19*, 391–414.

Kessler, W. S. (2006). The circulation of the eastern tropical Pacific: A review. *Progress in Oceanography, 69*, 181–217.

Kone, V., Machu, E., Penven, P., Andersen, V., Garcon, V., Fréon, P., et al. (2005). Modeling the primary and secondary productions of the southern Benguela upwelling system: A comparative study through two biogeochemical models. *Global Biogeochemical Cycles, 19*, GB4021. https://doi.org/10.1029/2004GB002427.

Lett, C., Penven, P., Ayón, P., & Fréon, P. (2007). Enrichment, concentration and retention processes in relation to anchovy (*Engraulis ringens*) eggs and larvae distributions in the northern Humboldt upwelling ecosystem. *Journal of Marine Systems, 64*, 189–200.

Marzloff, M., Shin, Y.-J., Tam, J., Travers, M., & Bertrand, A. (2009). Trophic structure of the Peruvian marine ecosystem in 2000–2006: Insights on the effects of management scenarios for the hake fishery using the IBM trophic model Osmose. *Journal of Marine Systems, 75*, 290–304.

Messié, M., & Chávez, F. (2011). Global modes of sea surface temperature variability in relation to regional climate indices. *Journal of Climate, 24*, 4314–4331.

Mogollón, R., & Calil, P. H. R. (2017). On the effects of ENSO on ocean biogeochemistry in the Northern Humboldt Current System (NHCS): A modeling study. *Journal of Marine Systems, 172*, 137–159. https://doi.org/10.1016/j.jmarsys.2017.03.011.

Mogollón, R., & Calil, P. H. R. (2018a). Counterintuitive effects of global warming-induced wind patterns on primary production in the Northern Humboldt Current System. *Global Change Biology*, 1–12. https://doi.org/10.1111/gcb.14171.

Mogollón, R., & Calil, P. H. R. (2018b). Modelling the mechanisms and drivers of the spatiotemporal variability of pCO2 and air–sea CO2 fluxes in the Northern Humboldt Current System. *Ocean Modelling, 132*, 61–72. https://doi.org/10.1016/j.ocemod.2018.10.005.

Moloney, C. L., Jarre, A., Arancibia, H., Bozec, Y.-M., Neira, S., Jean-Paul Roux, J.-P., et al. (2005). Comparing the Benguela and Humboldt marine upwelling ecosystems with indicators derived from inter-calibrated models. *ICES Journal of Marine Science, 62*, 493–502.

Montes, I., Colas, F., Capet, X., & Schneider, W. (2010). On the pathways of the equatorial subsurface currents in the eastern equatorial Pacific and their contributions to the Peru-Chile Undercurrent. *Journal of Geophysical Research, Oceans, 115*, C09003. https://doi.org/10.1029/2009JC005710.

Montes, I., Dewitte, B., Gutknecht, E., Paulmier, A., Dadou, I., Oschlies, A., et al. (2014). High resolution modeling of the Eastern Tropical Pacific oxygen minimum zone: Sensitivity to the tropical oceanic circulation. *Journal of Geophysical Research, Oceans, 119*(8), 5515–5532. https://doi.org/10.1002/2014JC009858.

Montes, I., Wolfgang, S., Colas, F., Blanke, B., & Echevin, V. (2011). Subsurface connections in the eastern tropical Pacific during La Niña 1999–2001 and El Niño 2002–2003. *Journal of Geophysical Research, 116*, C12022. https://doi.org/10.1029/2011JC007624.

Moore, J. K., Doney, S. C., & Lindsay, K. (2004). Upper ocean ecosystem dynamics and iron cycling in a global three-dimensional model. *Global Biogeochemical Cycles, 18*, GB4028. https://doi.org/10.1029/2004GB002220.

Muñoz, R. C., & Garreaud, R. D. (2005). Dynamics of the low-level jet off the west coast of subtropical South America. *Monthly Weather Review, 133*, 3661–3677. https://doi.org/10.1175/MWR3074.1.

Ñiquen, M., & M. Bouchón (2004). Impact of El Niño event on pelagic fisheries in Peruvian waters, Deep Sea Res. Part II, 51, 563–574. https://doi.org/10.1016/j.dsr2.2004.03.001.

Oerder, V., Colas, F., Echevin, V., Codron, F., Tam, J., & Belmadani, A. (2015). Peru-Chile upwelling dynamics under climate change. *Journal of Geophysical Research, Oceans, 120*, 1152–1172. https://doi.org/10.1002/2014JC010299.

Oliveros-Ramos, R., Verley, P., Echevin, V., & Shin, Y.-J. (2017). A sequential approach to calibrate ecosystem models with multiple time series data. *Progress in Oceanography, 151*, 227–244.

Oozeki, Y., Niquen, M., Takasuka, A., Ayon, P., Kuroda, H., Tam, J., et al. (2019). Synchronous multi-species alternations between the northern Humboldt and Kuroshio Current systems. *Deep-Sea Research Part II, 159*, 11–21.

Penven, P., Echevin, V., Pasapera, J., Colas, F., & Tam, J. (2005). Average circulation, seasonal cycle, and mesoscale dynamics of the Peru Current System: A modeling approach. *Journal of Geophysical Research, 110*, C10021. https://doi.org/10.1029/2005JC002945.

Resplandy, L., Levy, M., Bopp, L., Echevin, V., Pous, S., Sarma, V. V. S. S., et al. (2012). Controlling factors of the oxygen balance in the Arabian Sea's OMZ. *Biogeosciences, 9*, 5095–5109. https://doi.org/10.5194/bg-9-5095-2012.

Salvatteci, R., Field, D., Gutiérrez, D., Baumgartner, T., Ferreira, V., Ortlieb, L., et al. (2018). Multifarious anchovy and sardine regimes in the Humboldt Current System during the last 150 years. *Global Change Biology, 24*(3), 1055–1068.

Shchepetkin, A. F., & McWilliams, J. C. (1998). Quasi-monotone advection schemes based on explicit locally adaptive dissipation. *Monthly Weather Review, 126*, 1541–1580.

Shchepetkin, A. F., & McWilliams, J. C. (2005). The regional oceanic modeling system: A split-explicit, free-surface, topography-following-coordinate ocean model. *Ocean Modelling, 9*, 347–404.

Smith, A. D. M., Brown, C. J., Bulman, C. M., Fulton, E. A., Johnson, P., Kaplan, I. C., Lozano-Montes, H., Mackinson, S., Marzloff, M., Shannon, L. J., Shin, Y. J., & Tam, J. (2011). Impacts of fishing low-trophic level species on marine ecosystems. *Science 333*, 1147–1150. https://doi.org/10.1126/science.1209395.

Tam, J., Blaskovic, V., Goya, E., Bouchon, M., Taylor, M., Oliveros-Ramos, R., et al. (2010). Relación entre Anchovy y otros componentes del ecosistema. *Bol. Inst. Mar Perú., 25*, 31–37.

Tam, J., Jarre, A., Taylor, M., Wosnitza-Mendo, C., Blaskovic, V., Vargas, N., et al. (2009). Modelado de la merluza en su ecosistema con interacciones tróficas y forzantes ambientales. *Bol. Inst. Mar Perú., 24*, 27–32.

Tam, J., Taylor, M. H., Blaskovic, V., Espinoza, P., Ballón, R. M., Díaz, E., et al. (2008). Trophic modeling of the Northern Humboldt Current Ecosystem, part I: comparing trophic trophic linkages under La Niña and El Niño conditions. *Progress in Oceanography, 79*, 352–365.

Tarazona, J., & Arntz, W. (2001). The Peruvian Coastal Upwelling System. *Coastal Marine Ecosystems of Latin America, 144*, 229–244. https://doi.org/10.1007/978-3-662-04482-7_17.

Taylor, M. H., Tam, J., Blaskovic, V., Espinoza, P., Ballón, R. M., Wosnitza-Mendo, C., et al. (2008). Trophic modeling of the Northern Humboldt Current Ecosystem, Part II: Elucidating ecosystem dynamics from 1995 to 2004 with a focus on the impact of ENSO. *Progress in Oceanography, 79*, 366–378.

Thomas, A. C., Carr M. E., & Strub, P. T. (2001). Chlorophyll variability in eastern boundary currents, Geophys. *Research Letters, 28*(18), 3421–3424. https://doi.org/10.1029/2001GL013368.

Tovar, H., & Cabrera, D. (1985). Las aves guaneras y el fenómeno "El Niño", in El fenómeno "El Niño" y su impacto en la fauna marina. W. F. Arntz, A. Landa, & J. Tarazona, Bol. Inst. Mar Perú, extraordinary volume, (pp. 181–186).

Trasmonte, G., & Silva, Y. (2008). Evento La Niña: Propuesta de definición y clasificación según las anomalías de temperatura de la superficie del mar en el área Niño 1+2, *Informe Instituto del Mar del Perú 35*(3), 199–207.

Vergara, O., Dewitte, B., Montes, I., Garçon, V., Ramos, M., Paulmier, A., & Pizarro, O. (2016). Seasonal variability of the oxygen minimum zone off Peru in a high-resolution regional coupled model, *Biogeosciences, 13*, 4389–4410. https://doi.org/10.5194/bg-13-4389-2016

Walsh, J., & Dugdale, J. (1971). A simulation model of the nitrogen flow in the Peruvian upwelling system. *Investigacion Pesquera, 35*, 309–330.

Walsh, J. J. (1981). A carbon budget for overfishing off Peru. *Nature, 290*, 300–304.

Xu, Y., Chai, F., Rose, K. A., Niquen, M., & Chavez, F. P. (2013). Environmental influences on the interannual variation and spatial distribution of Peruvian anchovy (Engraulis ringens) population dynamics from 1991 to 2007: A three-dimensional modeling study. *Ecological Modelling, 264*, 64–82.

Yang, S., Gruber, N., Long, M. C., & Vogt, M. (2017). ENSO driven variability of denitrification and suboxia in the Eastern Tropical Pacific Ocean. *Global Biogeochemical Cycles, 31*, 1470–1487. https://doi.org/10.1002/2016GB005596.

Yonss, J. S., Dietze, H., & Oschlies, A. (2017). Linking diverse nutrient patterns to different water masses within anticyclonic eddies in the upwelling system off Peru. *Biogeosciences, 14*, 1349–1364. https://doi.org/10.5194/bg-14-1349-2017.

Using Ecosystem Models to Evaluate Stock Recovery in Two Hake Species from Chile

Sergio Neira and Hugo Arancibia

1 Introduction

Fisheries are one of the most important human activities in the marine ecosystems, providing food, jobs, income, and social wellbeing (FAO, 1995). However, several world fisheries have been—or continue to be—fished beyond sustainable levels. As a result, several stocks are experiencing overfishing and collapse (Cochrane, 2002; FAO, 1997, 2016). These processes can generate profound and sustained ecological and socio-economic loss, since stock recovery after collapse is rather slow and sometimes nil (e.g., Atlantic cod in Canadian EEZ; Hutchings & Rangeley, 2011). The usual suspects behind the lack of recovery are excessive fishing, unfavorable environmental conditions, life history characteristics that slow down recovery, and poor management performance (Hammer et al., 2010).

Fisheries have been traditionally managed using single-species approaches alone using the maximum sustainable yield (MSY, Schaefer, 1991) as sustainability goal. MSY corresponds to the maximum yield that can be continuously obtained from a stock and it has been adopted by several intergovernmental organizations and

S. Neira (✉)
Departamento de Oceanografía, Universidad de Concepción, Concepción, Chile

Centro de Investigación Oceanográfica COPAS Sur-Austral, Universidad de, Concepción, Chile

Programa de Doctorado en Ciencias con mención en Manejo de Recursos Acuáticos Renovables (MaReA), Facultad de Ciencias Naturales y Oceanográficas, Universidad de Concepción, Concepción, Chile
e-mail: seneira@udec.cl

H. Arancibia
Departamento de Oceanografía, Universidad de Concepción, Concepción, Chile

Programa de Doctorado en Ciencias con mención en Manejo de Recursos Acuáticos Renovables (MaReA), Facultad de Ciencias Naturales y Oceanográficas, Universidad de Concepción, Concepción, Chile

M. Ortiz, F. Jordán (eds.), *Marine Coastal Ecosystems Modelling and Conservation*, https://doi.org/10.1007/978-3-030-58211-1_4

international agreements such the United Nations Convention on the Law of the Sea (United Nations, 1997), the United Nations Fish Stocks Agreement (1995), the Code of Conduct for Responsible Fisheries (1995), and the United Nations Sustainable Development Summit (United Nations, 2002), among others. The majority of these agreements advise maintaining or restoring stock biomass to the level producing MSY (Urrutia et al., 2015). However, fish stocks are not isolated from their surrounding ecosystem and the populations dynamics of fish is tightly associated with other living (e.g., prey, predators, competitors) and non-living (e.g., physical and chemical factors) ecosystem components. Unfortunately, the standard scientific advice and decision-making processes in fisheries management are based on single-species assessments that do not incorporate multispecies, food web, and whole ecosystem (including socio-economic) dimensions. This is against the modern view that conceptualizes fisheries as socio-ecological systems, and several international agreements call for a rapid movement towards a more holistic ecosystem approach to fisheries management (FAO, 2003).

This new ecosystem management approach requires the application of scientific methods and tools to advise managers and decision-makers on management actions that consider the wider range of societal objectives and the interactions in the ecosystem. In this context, ecosystem models (i.e., models that represent a wider range of technological and ecological processes affecting the species in the ecosystem) ranging from multispecies to whole ecosystems are key tools for providing this scientific information and advice (FAO, 2008). In particular, The *Ecopath* with *Ecosim* software and model (Christensen & Pauly, 1992; Walters et al., 1997) is widely used tool in the field of ecosystem modelling. This approach is also regarded as important tool to support the ecosystem approach to fisheries management, based on its flexibility to address the widest range of research and management questions (Plagányi, 2007).

In year 2013, the Chilean Fisheries and Aquaculture Law was amended declaring conservation and sustainable use of living aquatic resources as main management goal. In addition, the Law mandates the application of both the ecosystem approach and the precautionary approach into management actions, as well as the safeguard of marine ecosystems. This Law defines the ecosystem approach as the one that considers the interrelations among predominant species in a specific area, and also mandates that, in each fishery, annual fishing quotas should maintain or recover stock biomass to the biomass generating MSY (B_{MSY}). Nevertheless, all fishing quotas in Chile are estimated using single-species models and indicators (mostly B_{MSY} and the fishing mortality leading to MSY). For overexploited stocks (i.e., spawning biomass < target spawning biomass), management plans are mandated, which need to include a recovery program.

Chilean hake (*Merluccius gayi*) and Southern hake (*M. australis*) are the most important demersal fish species supporting important fisheries in central and southern Chile, respectively (Arancibia et al., 2016; Fig. 1). The landings of both species are directed to manufacture products for direct human consumption with final destination in Chilean and international markets (Arancibia, 2015; Gatica et al., 2015; Arancibia et al., 2016). Both fisheries support jobs and income for small-scale and

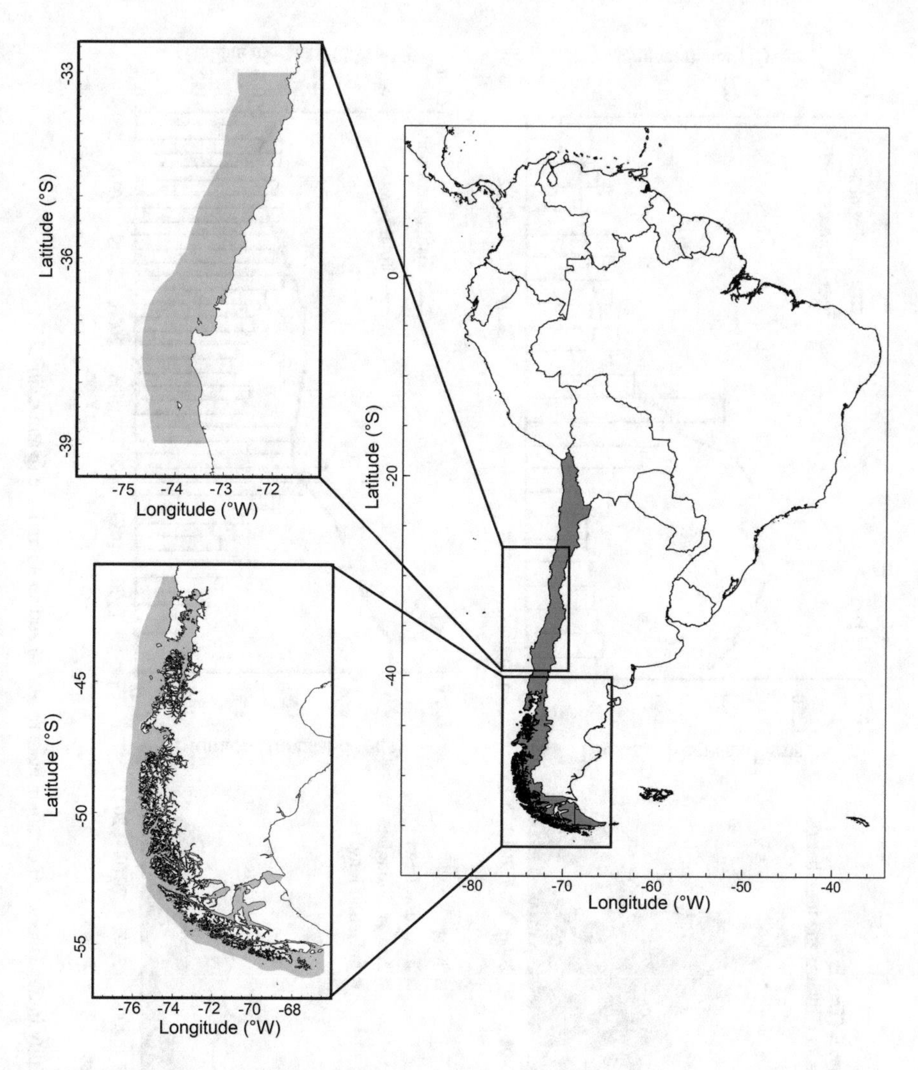

Fig. 1 Study areas corresponding to the main operation areas of the fisheries targeting common hake and southern hake

large-scale fleets, as well as in production plants located mainly in the ports of Talcahuano (Chilean hake) and Puerto Chacabuco (Southern hake). These fisheries have been either collapsed (common hake) or overexploited (Southern hake) for several years, with little signs of recovery (Fig. 2). Therefore, these fisheries need recovery plans and programs.

Chilean hake is a demersal fish species inhabiting the upwelling ecosystem off central Chile (32°30′–37°30′S) and it is distributed on the continental shelf and slope from 50 to 400 m deep (Gatica et al., 2015). The common hake fishery comprises a bottom-trawl industrial fleet (large vessels) and an artisanal fleet (boats)

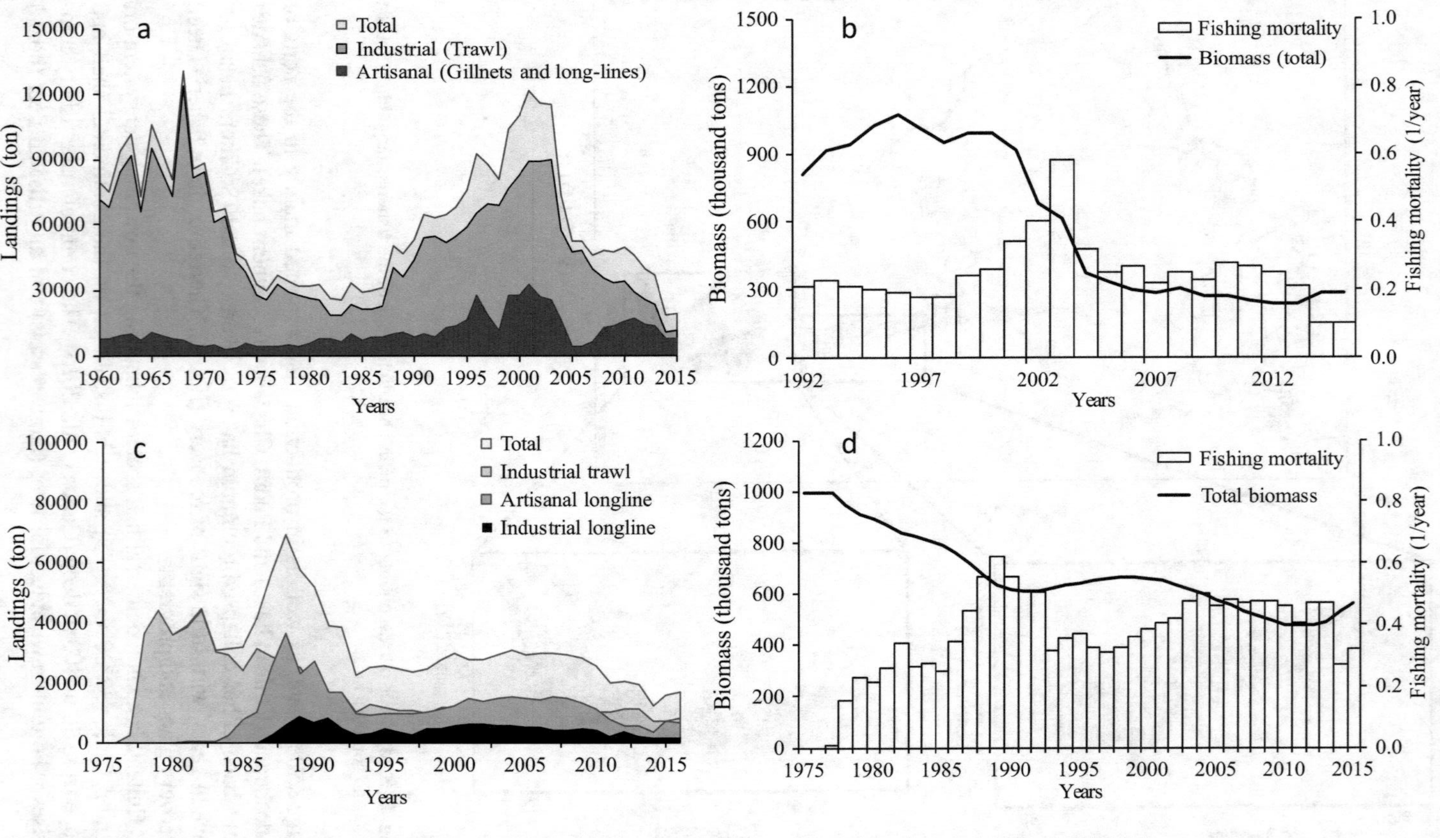

Fig. 2 Landings, biomass, and fishing mortality in the fisheries of common hake (**a** and **b**) and southern hake (**c** and **d**) in Chile

operating with long-lines and gillnets. The fishery began in the 1940s and has been ever since oriented to direct human consumption, mainly as fresh and frozen fillets. This fishery represents one of the most valuable fisheries in Chile in terms of income and jobs (Arancibia & Neira, 2005; Gatica et al., 2015). Common hake plays an important role as predator, with consistent predation on small pelagic fish (mainly common sardine *Strangomera bentincki* and anchovy *Engraulis ringens*), lantern fish, and crustaceans (such red squat lobster *Pleuroncodes monodon* and yellow squat lobster *Cervimunida johnii*). Common hake also exhibits a strong cannibalistic behavior, especially on juveniles smaller than 18 cm of total length (Jurado-Molina et al., 2006; Cubillos et al., 2007). Some predators such sea lions *Otaria flavescens* (Husktadt and Antezana, 2003) and Jumbo squid *Dosidicus gigas* (Arancibia & Neira, 2008; Ibáñez et al., 2008) can predate on Chilean hake. The stock size has been around 0.2*B_{MSY} since early 2000s, and accordingly the status is overfished.

In Southern Chile (from 41°28,6'S to 57°S) there is a fishery targeting southern hake and other demersal species such hoki (*Macruronus magellanicus*), Southern blue whiting (*Micromesistius australis*), kingklip (*Genypterus blacodes*), and skates (species of the Family Rajidae). These species exhibit strong trophic interactions and most of them are piscivores predators as adults (Payá, 1992; Arancibia et al., 2010; Neira et al., 2012; Jurado-Molina et al., 2016). The fishery started in late 1970s and reached maturity in 1980s and 1990s with well-developed large-scale and small-scale fleets. The small-scale fleet includes boats operating in the inner sea (0–5 nautical miles from the shore line), using long-lines. The large-scale fleet encompasses trawlers and long-liner vessels that operate in open sea (5–80 nautical miles). Each stock is targeted by both kind of fleets and, therefore, there are technical interactions. In year 2018, the stocks of skates, southern hake, southern blue whiting, and kingklip were considered as overexploited (i.e., $B_{\text{Limit}} < B_{2018} < B_{\text{MSY}}$), while the stock of hoki was considered collapsed (i.e., $B_{2018} < B_{\text{Limit}}$).

Considering the trophic interactions of common hake and Southern hake as well as the technical interactions among fleets targeting these species and their prey and predators, there is concern that a management system based on single-species biological reference points (MSY) may not be precautionary nor consider the ecosystem approach requested by both the Chilean Fisheries and Aquaculture Law and global agreements. Therefore, in this chapter we evaluate stock recovery in two species of hakes from Chile by means of ecological modelling, by explicitly considering the trade-offs among ecological, social, and economic objectives in these ecosystems.

2 Study Area and Modelling Framework

We used two ecosystem models representing the main fishing grounds for common hake (central Chile from 33 to 39°S, upwelling system) and southern hake (southern Chile from 41 to 57°S, non-upwelling system). Models representing the central and southern zones of Chile were based on previous models built by Neira & Arancibia (2004) and Arancibia et al. (2010), respectively. Models were built using the Ecopath with Ecosim software model, EwE (Christensen & Pauly, 1992; Walters et al., 1997).

The input data used to parameterize the models corresponded to those of year 2015, since this period allowed to collect fishery, ecological, and socio-economic data for the main fisheries in each system. We consider the main artisanal and industrial fleets in each system and for each fleet the model required data on the magnitude and price of landings/catch, the number of jobs, and the direct and indirect costs. The data were collected by Arancibia et al. (2016) and are summarized in Table 1.

Following Arancibia & Neira (2008), the model representing the central Chile system considered the following fishing fleets: hake artisanal (long-lines and gill-nets), hake industrial (trawl), artisanal purse-seine (targeting common sardine and anchovy), industrial purse-seine (targeting Horse mackerel, anchovy, and common sardine), crustacean trawlers (targeting red and yellow squat lobsters), and an artisanal fleet targeting Jumbo squid. On the other hand, the model representing Southern Chile considered the following fleets: hake artisanal (long-lines), hake industrial (trawl), hoki industrial (trawl), Southern blue whiting (trawl), other demersal fish artisanal (targeting kingklip and skates), and artisanal purse-seiners. Following Arancibia et al. (2010) and Jurado-Molina et al. (2016), the model representing the Southern zone of Chile comprises inner and open waters from 41°28,6'S to 57°00'S, and from the coast line to 80 nautical miles westward. The total surface area is about 290 thousand km^2 (Fig. 1).

2.1 Mathematical Model and Functional Groups

The models included all trophic levels in each ecosystem, ranging from primary producers to top predators. Models focused on the dynamics of the main target species for the fisheries in the study areas (common hake and Southern hake) as well as their prey and predators. We used the EwE (Christensen & Pauly, 1992; Walters et al., 1997) as modelling platform. EwE is based on two main assumptions: (1) the production of each functional group in the model is balanced by predation, exports, and mortalities, and (2) the consumption of each group is balanced by production, respiration, and unassimilated food (i.e., mass-balance). The mathematical expressions are

Table 1 Socio-economic data used as input in simulations evaluating management scenarios

Species	Fleet	Price (US$/kg)	Jobs	Landings
Common hake	Artisanal	1.00	8752	9028
	Industrial	4.00	789	13,452
Southern hake	Artisanal	1.86	12,148	9683
	Industrial	13.00	2552	6456

$$P_i = Y_i + B_i M2_i + E_i + \mathrm{BA}_i + M0_i \tag{1}$$

where P_i is the total production of group i; Y_i is the catch of i; $M2_i$ is predation mortality of i; B_i is total biomass of i; E_i is the migration rate; BA_i is the accumulation of biomass of i; $M0_i = P_i(1 - \mathrm{EE}_i)$ is the other mortality rate (those independent from predation and catches). Equation (1) can be re-arranged as:

$$B_i (P/B)_i - \sum_{j=1}^{n} B_j \left(\frac{Q}{B}\right)_i \mathrm{DC}_{ji} - \left(\frac{P}{B}\right)_i B_i (1 - \mathrm{EE}_i) - Y_i - E_i - \mathrm{BA}_i = 0 \tag{2}$$

where $(P/B)_i$ is production to biomass rate, $(Q/B)_i$ is consumption to biomass rate, and DC_{ji} is the fraction of prey i in the diet of predator j.

The mass-balance for each group is given by the following equation:

$$Q = P + R + U \tag{3}$$

where Q is consumption, P is production, R is respiration, and U is unassimilated food.

The temporal dynamics of the biomass of each functional group i is defined by the following equation:

$$\frac{dB_i}{dt} = g_i \sum_{j=1}^{n} c_{ji}\left(B_i, B_j\right) - M0_i B_i - F_i B_i - \sum_{j=1}^{n} c_{ij}\left(B_i, B_j\right) \tag{4}$$

where $\frac{dB_i}{dt}$ is the rate of change in the biomass of group i in time interval t; g_i is the net growth efficiency (P/Q) for group i; $c_{ji}(B_i, B_j)$ is the function that predicts the consumption of prey i by predator j; $M0_i$ is other mortalities of i; and F_i is the fishing mortality rate of i. This model allows temporal simulations to evaluate the effects of forcing functions on the biomass of one or more functional groups.

The consumption $c_{ji}(B_i, B_j)$ is predicted using the concept of "foraging arena" as follows. Each B_i is split into two parts, the first one is the vulnerable biomass (V_{ij}) and the second one is invulnerable biomass ($B_i - V_{ij}$). In this sense, a transference rate (v_{ij}) among V_{ij} and $B_i - V_{ij}$ is defined to represent the maximum predation rate that predator j can exert on prey i. The equation is

$$\frac{dV_{ij}}{dt} = v_{ij}\left(B_i - V_{ij}\right) - v_{ij} V_{ij} - a_{ij} V_{ij} B_j \tag{5}$$

where a_{ij} is the effective search rate of predator j on prey i, and v_{ij} is the vulnerability that represents the degree of predation mortality in a prey caused by the increase in the biomass of a predator.

2.2 Optimization Routines

We run optimization routines contained in EwE to obtain fleet-specific fishing mortality rates that maximize in each model the economic (income), social (jobs), mandated rebuilding (increase in the biomass of a particular stock to a desired level, in this case B_{MSY}), and ecological (conservation of ecosystem's trophic structure) objectives over 20 years. These optimizations used the Davidon–Fletcher–Powell approach (Christensen et al., 2005), which is a non-linear searching procedure that optimizes an objective (searches maximum and minimum values in a complex function) by changing fishing mortality rates in the fleets included in each model. The routine maximized the following objective function:

$$F_{\text{Total}} = \left(w_{\text{economic}} * \text{income}\right) + \left(w_{\text{social}} * \text{jobs}\right) + \left(w_{\text{rebuilding}} * \text{biomass}_i\right) + \left(w_{\text{ecologic}} * \text{ecosystem strucure}\right)$$

where F_{Total} are the fleet-specific fishing mortality rates that maximize a management objective, w_{economic}, w_{social}, $w_{\text{rebuilding}}$, and $w_{\text{ecosystem}}$ are relative weights that the modeler includes depending on the objective to be maximized, *income* corresponds to the value of the catch in each fleet subtracting the costs of fishing (fixed and variable), *jobs* corresponds to the number of jobs supported by each fleet, and *ecosystem structure* corresponds to the inverse of the P/B ratio that is a measure of ecosystem maturity (*sensu* Odum, 1969).

To emulate the decision-making process, several management scenarios were developed. In this process, we assigned a weighting factor to every index (economic, jobs, ecosystem structure, biomass), depending on which objective was preferred to be accomplished. The scenarios were named as follows: "economic scenario" ($), which maximizes the total income obtained from all fleets in each ecosystem; "social scenario" (S), which maximizes the total jobs obtained from all fleets in each ecosystem; "ecosystem structure scenario" (E), which maximizes the biomass of big-sized species in the ecosystem; and "mandated recovery scenario" (R), which restores biomass of species *i* to a target biomass level (B_{MSY}). We assigned a weighting factor of 1 to the criterion to be maximized and 0 to the criteria that were not maximized. For example, to accomplish the objective of maximize the total income obtained from all fleets in each ecosystem, scenario $, we used the following weighting factors: $ = 1; *S* = 0; *E* = 0; *R* = 0.

Considering that rebuilding to B_{MSY} in common hake and Southern hake is mandatory in the Chilean Law, we did not explore maximization of individual objectives (i.e., economic, social, ecosystem), but they were analyzed in combination with the recovery objective. Therefore, the combinations analyzed were: economic and recovery ($ + R: 1,0,1,0); social and recovery (S + R: 0,1,1,0), mandated recovery of the stocks of common hake or southern hake (R: 0,0,1,0); ecosystem and recovery (E + R: 0,0,1,1). We simulated a fifth scenario considering the simultaneous maximization of all four objectives or "ecosystem and social wellbeing" (*W* = 1,1,1,1). Finally, we simulated a *status quo* scenario (current situation) in which no objective was maximized, i.e., (SQ: 0,0,0,0).

We considered the net income of each fleet as economic sustainability index. This indicator was calculated as total utilities subtracting costs (fixed and variables) in each fleet. The economic objective implied to maximize the total net income obtained from all fleets in each system. The costs and income of other fleets in the central-south zone were obtained from Arancibia & Neira (2005), while those of the southern zone were obtained from Arancibia et al. (2016). Landings of the main fishing resources in both systems were obtained from the official annual statistics from the Chilean Fisheries Service (www.sernapesca.cl). The off-vessel value of the landings was obtained by direct interviews with key actors in each fishery (stakeholders, fishers, managers, others).

We considered the number of jobs sustained by each fleet as social sustainability index. This indicator was calculated as the jobs/landings ratio (J/C). This emulates the social objective aiming to maximize the social benefits (jobs) sustained by fishing activities. The social objective corresponded to the maximization of the total job sustained by the different fisheries in each system. The social index was constructed using job and landings values informed by Arancibia & Neira (2008), Arancibia et al. (2016), and of the annual reports of the Chilean Fisheries Service. Likewise, we considered also the ecologic sustainability index as the inverse of the production/biomass ratio (P/B) of the functional groups included in each model. These values were obtained from Neira & Arancibia (2004), and Arancibia et al. (2010). The $(P/B)^{-1}$ ratio corresponds to a measure of the potential growth/recovery of each functional group. Following Odum's ecological theory (Odum, 1969, 1971, 1985), mature and stable ecosystems are characterized by sustaining high biomass big-sized and long-lived organisms. Therefore, the ecological objective implies to maintain or to improve ecosystem's structure and function and the ecological scenario (E) corresponds to maximize the biomass of big-size and long-lived functional groups (and by doing so, increasing the biomass of their prey).

The mandated rebuilding objective (R), implied finding the fishing mortality rates that, based on the productivity of the target species, allow biomass recovery of common hake and southern hake to B_{MSY}. The productivity of the functional groups in each model was obtained from Neira & Arancibia (2004) and Arancibia et al. (2010). We considered a minimum biomass recovery of 1.5 and 2 times the current spawning biomass in the case of southern hake and common hake, respectively. The above recovery targets were set considering that, in year 2015, the spawning biomass in the Southern hake stock was at approximately 70% of B_{MSY} (Quiroz et al., 2015), while the spawning biomass in common hake was about 20% of B_{MSY} (Tascheri et al., 2014).

The socio-economic data (income, costs, and jobs) used to explore optimizations are presented in Tables 1 and 2. For each scenario, we calculated the relative change of the following indices: total income, total jobs, the biomass of big-sized species in the ecosystem, and the biomass of hake stocks relative to B_{MSY}. These changes were compared with the value of each index in the *status quo* scenario. Obviously, the maximization of any individual objective (or combination of objectives) resulted in changes in the above indices in relation with the *status quo* scenario. We considered plausible management scenarios those presenting simultaneous increase in all four

Table 2 Cost and profit data used as input in simulations, fishery of common hake. Data is expressed in percentage, so profit + costs = 100%

Zone	Fleet name	Fixed cost (%)	Effort related cost (%)	Sailing related cost (%)	Profit (%)	Total value (%)
Central Chile	Industrial hake	18	26	26	30	100
	Artisanal hake	5	22.5	22.5	50	100
Southern hake	Industrial hake	20	25	25	30	100
	Artisanal hake	5	22.5	22.5	50	100

Table 3 Relative change in socio-ecological indicators under fisheries management scenarios simulated in the model representing the central Southern zone of Chile

	Indicators			
Scenario	Economic	Social	Rebuilding	Ecologic
Status quo	1.00	1.00	1.00	1.00
Economic + mandated rebuilding	1.24	1.11	1.12	1.14
Social + mandated rebuilding	−5.35	1.40	1.57	1.02
Mandated rebuilding	−5.22	1.21	1.55	1.01
Ecosystem + mandated rebuilding	0.76	1.54	1.03	5.32
All	−0.07	1.61	1.17	5.13

indices, i.e., $ > 1; S > 1; E > 1 y R > 1. These plausible scenarios were kept for further analysis.

We used Eq. (4) to analyze the trajectories of the biomass of selected functional groups in each model, under the combination of fleet-specific fishing mortality coefficients resulting from plausible scenarios. Under each scenario selected we analyzed whether the minimum recovery biomass was reached in both hake species, and whether the biomass of other functional groups increased or declined compared the current situation. Simulations were run for 20 years, which allowed to evaluate recovery time for common hake and Southern hake, and changes in the biomass of other functional groups (including other target species for the fisheries) and the whole food web.

3 Modelling Outcomes

In Table 3 we present the relative change in the economic, social, ecological, and rebuilding indices that resulted from the maximization of management objectives in the model representing the central Chile zone. The biomass of common hake showed the larger increase in the scenario aiming mandated rebuilding and in the scenario combining maximization of social and mandated rebuilding. However, in these scenarios, the increase in common hake biomass was around 60% and therefore the rebuilding target for common hake (set at 2 times biomass level in year 2015) was not met. Other combinations resulted in a moderate increase in common hake biomass of 8% (economic + mandated rebuilding) and an increase of 5% (ecology + mandated rebuilding), while others resulted in decline in common hake biomass of 2% (social + mandated rebuilding) and 47% (all objectives together). The mandated rebuilding objective for common hake resulted in strong decline in the economic (total income) and a 25% increase in the social (jobs) indices. In Table 4 we present the fleet-specific fishing mortality rates obtained under each management scenario simulated. The mandated rebuilding biomass of common hake resulted in a strong decline in the fishing mortality exerted by the industrial fleet targeting common hake, and fleets targeting small pelagics fish that are preyed upon by common hake. These fleets are industrial and artisanal purse-seiners targeting anchovy and common sardine. The mandated rebuilding scenario indicated an increase in the artisanal fleet targeting common hake and squat lobsters. All scenarios indicated a moderate to strong increase in the fishing mortality exerted by the fleet targeting Jumbo squid.

In Table 5 we present the relative changes in the economic, social, ecological, and mandated rebuilding indices resulting from the maximization of objectives in the model representing the southern-austral zone of Chile. The biomass of Southern hake increased in all scenarios in about 30%, except the scenario that combined the maximization of ecosystem structure and mandated rebuilding that resulted in a 24% biomass decline. Therefore, the biomass recovery target of 50% for biomass of Southern hake was not met. The mandated rebuilding of southern hake resulted in a 24% decline in the economic (total income) indicator and a 22% decline in the social indicator (jobs). In Table 6 we present the fleet-specific fishing mortality rates obtained under each optimization conducted. The mandated rebuilding in Southern hake resulted in a strong decline in the fishing mortality related to the fleet targeting other demersal fish, the fleet targeting it should read Southern blue whiting, and the

Table 4 Fishing mortality rates resulting from fisheries management scenarios simulated in the model representing the central Southern zone of Chile

	Fleets					
Scenario	Hake artisanal	Hake industrial	Industrial purse-seine	Artisanal purse-seine	Squat lobsters	Jumbo squid artisanal
Status quo	1.00	1.00	1.00	1.00	1.00	1.00
Economic + mandated rebuilding	0.66	2.30	1.54	0.10	3.70	1.51
Social + mandated rebuilding	3.97	0.17	0.52	0.18	0.21	10.00
Mandated rebuilding	2.91	0.11	0.83	0.28	2.48	10.00
Ecosystem + mandated rebuilding	0.10	1.00	0.12	0.60	0.36	0.35
All	0.10	0.94	3.27	1.12	0.61	2.51

Table 5 Relative change in socio-ecological indicators under fisheries management scenarios simulated in the model representing the south-austral zone of Chile

	Objective			
Scenario	Economic	Social	Rebuilding	Ecologic
Status quo	1.00	1.00	1.00	1.00
Economic + mandated rebuilding	11.98	5.35	1.34	1.65
Social + mandated rebuilding	2.12	8.68	1.28	1.49
Mandated rebuilding	0.76	0.78	1.25	1.75
Ecosystem + mandated rebuilding	1.48	1.61	0.76	6.87
All	5.13	8.69	1.31	6.16

Table 6 Fishing mortality rates resulting from fisheries management scenarios simulated in the model representing the south-austral zone of Chile

	Fleets					
Scenario	Hake industrial	Hake artisanal	Industrial purse-seine	Other demersal artisanal	Trawlers Southern blue whiting	Artisanal purse-seiners
Status quo	1.00	1.00	1.00	1.00	1.00	1.00
Economic + mandated rebuilding	6.32	0.10	9.19	0.50	0.97	1.31
Social + mandated rebuilding	0.82	9.21	10.00	0.91	3.85	0.10
Mandated rebuilding	0.19	0.23	1.42	0.38	0.52	0.10
Ecosystem + mandated rebuilding	1.79	0.90	0.25	0.10	1.99	1.50
All	2.31	6.54	10.00	0.12	2.17	10.00

fleet targeting small pelagic fish. This scenario also resulted in an increase in the fleet targeting hoki.

In Fig. 3 we show the biomass trajectory for the main functional groups under the fleet-specific fishing mortality rates that optimized selected scenarios in the model representing the central Chile ecosystem. The fishing mortalities related to the mandated rebuilding scenario (Table 3) allowed a moderate increase in the biomass of adult hake (hake II) as well as juvenile hake (hake II), while the biomass of Jumbo squid collapsed in this scenario (Fig. 3a). Under the social + recovery scenario, all fishing mortalities strongly declined (Table 4) allowing the increase of common hake prey (e.g., anchovy and common sardine). In this scenario, the biomass of squat lobsters (macrobenthos) slightly declined and the biomass Jumbo squid collapsed (Fig. 3b).

In Fig. 4 we show the trajectory of main functional groups under fleet-specific fishing mortality rates that optimized selected scenarios in the southern zone of Chile. The fishing mortalities related to the mandated rebuilding scenario (Table 6) allowed an increase in the biomass of southern hake and hoki juveniles, while the biomass of other pelagic fish, Southern blue whiting, and adult of hoki showed a decreasing trend (Fig. 4a). Under the economic + recovery scenario, the biomass of

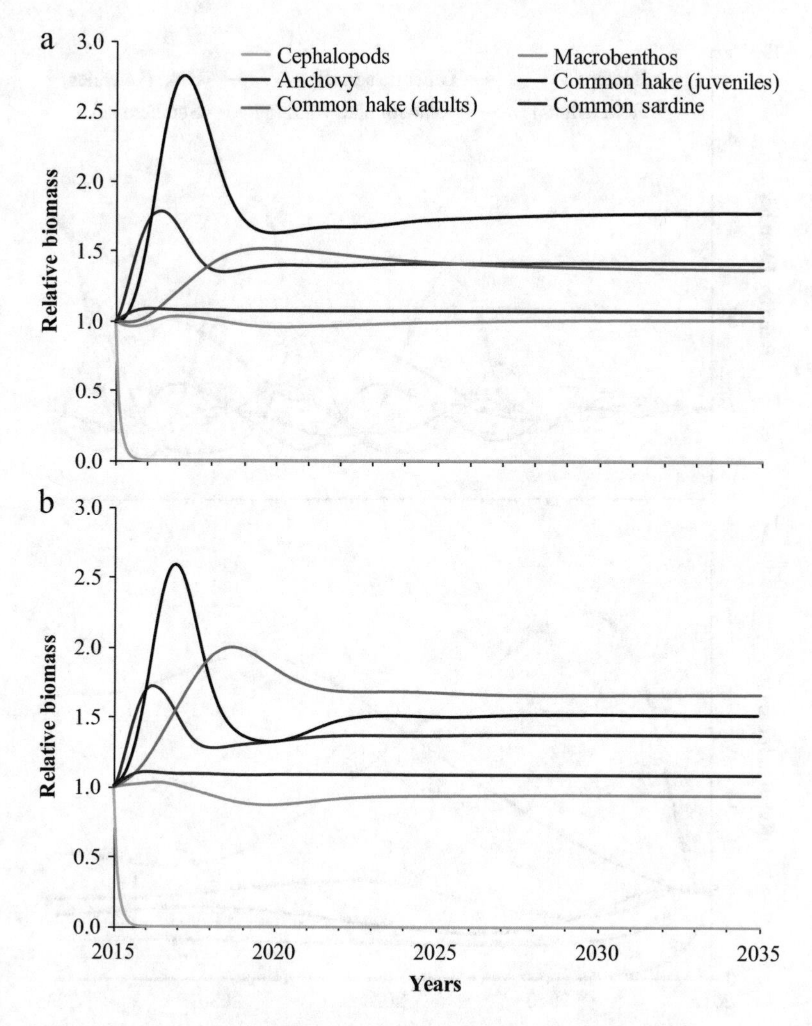

Fig. 3 Biomass trends in important stocks under the scenario of mandated rebuilding of Chilean hake to the biomass level producing the maximum sustainable yield (**a**) and the scenario combining the maximization of the social (jobs) and mandated rebuilding of common hake biomass (**b**)

Southern hake and hoki juveniles increased, while the biomass of Southern blue whiting declined and the biomass of hoki adults collapsed (Fig. 4b).

4 Fisheries and Socio-Economical Constrains

In the stocks of common hake and Southern hake, the objectives of economic and social sustainability are not compatible with the objectives of stock recovery and ecological sustainability. Results showed that it is not possible to recover these

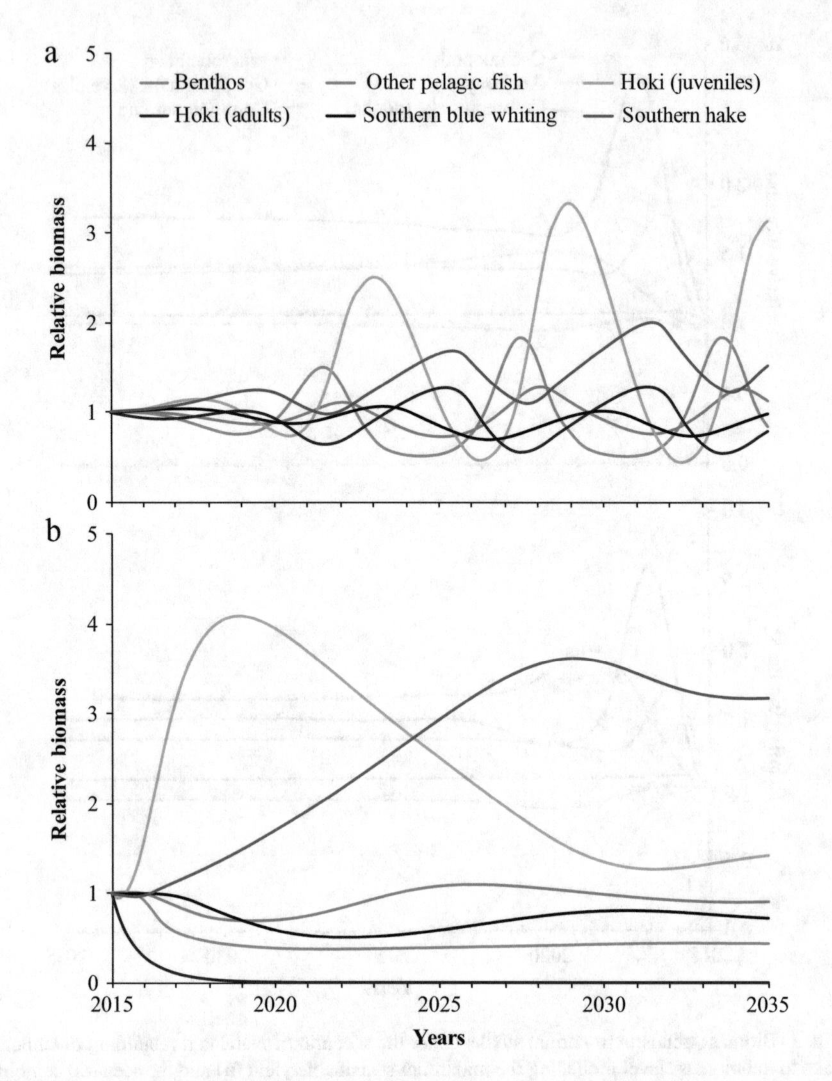

Fig. 4 Biomass trends in important stocks under the scenario of mandated recovery of Southern hake to the biomass level producing the maximum sustainable yield (**a**) and the scenario combining the maximization of the economic (maximization of income) and mandated rebuilding of southern hake (**b**)

stocks and their ecosystems without affecting the socio-economic wellbeing (income and jobs), and vice versa it is not possible to increase jobs and income without compromising stock recovery and ecosystem wellbeing.

The optimization routine compensated the decline in jobs and income resulting from forcing the recovery of common hake and Southern hake by distributing fishing effort towards other fleets operating in each system. In the case of common hake, the effort increased heavily in the fleet targeting jumbo squid (Table 4).

However, this change is not sustainable since it leads to stock collapse in jumbo squid (Fig. 3). In the case of mandated rebuilding in Southern hake, effort increased in the trawling fleet targeting hoki (Table 6). Obviously, this change is not realistic, since the fishery of hoki is currently overfished and an increase in fishing mortality/ effort is not advisable.

In the case of common hake, and in spite of a strong decrease in fishing mortality, it was not possible to accomplish the rebuilding target of doubling current biomass (Table 4). In this sense, results indicated that in order to recovery stock biomass in common hake it is also necessary to decrease fishing mortality in its main prey (anchovy, common sardine). These model results are somehow intuitive considering that the diet of common hake adults (i.e., individuals >37 cm total length) is based on small pelagic fish (common sardine and anchovy), benthic crustaceans (red squat lobster, and yellow squat lobster), and cannibalism on small hake individuals (Neira et al., 2004; Cubillos et al., 2007).

In the case of southern hake, the rebuilding target was not met either. The rebuilding seems more likely in this species, but negative impacts on the economic and social objectives are to be expected (Table 5). In addition, to rebuild Southern hake stock it is necessary to heavily increase fishing effort in the trawling fleet targeting hoki adults (Table 6). One explanation for this result is that the group hoki juveniles is the main prey for southern hake (~80% in weight), but also for hoki adults (Payá, 1992; Arancibia et al., 2010; Neira et al., 2012; Jurado-Molina et al., 2016). Therefore, increasing effort on hoki adults releases biomass of hoki juveniles that is then more available for southern hake. However, these model outcomes are not quite realistic since the dynamics of hoki juveniles is not independent from the dynamics of hoki adults. Moreover, a sustained increase of juveniles is not expected from a decline in adults. Nevertheless, model results indicate that stock recovery in southern hake would require more biomass of its main prey (hoki juveniles). The assessments of hoki indicate a decline in the abundance of recruits in the last decades (Alarcón & Zuleta, 2013; Payá et al., 2014), and therefore increase in hoki juveniles (prey) is not expected in the short term. Therefore, if the recovery of southern hake depends on the recruitment of hoki besides fishing mortality, we project that the stock recovery of southern hake is not likely in the medium to long term.

The ecosystem approach to fisheries has been treated mostly as an ecological question. However, fisheries are socio-ecological systems, and recovery does not depend only on the ecological factors that affect stock abundance (e.g., predator–prey interactions, influence of environment on stock dynamics). Instead, recovery (and ultimately sustainability) heavily depends on the capacity of the management system to adjust fishing mortality towards appropriate levels. In addition, there is need to advance our knowledge on the socio-ecological system and the behavior of users and other stakeholders. The *Ecopath with Ecosim* approach, as well as other modelling platforms, is a helpful tool to advise or inform managers, decision-makers, and other stakeholders on the likely impacts that fisheries policies such as mandated rebuilding may have on the social and ecosystem wellbeing. We envision an increasing use of ecosystem models to support fisheries management in order to secure sustainability and conservation of Latin American marine coastal fisheries and their ecosystems.

5 Conclusions

1. The objectives of economic and social sustainability are not compatible with the objectives of mandated rebuilding in common hake and in southern hake. This means that it is not possible to recover the biomass of common hake and southern hake (and their ecosystem) without affecting income, jobs and vice versa.
2. The stock dynamics and recovery in common hake and Southern hake may be strongly influenced by trophic factors (predator–prey interactions), which are not considered in single-species models currently used to determine the biological acceptable quotas for each species and to evaluate recovery scenarios in both species.

Acknowledgements Sergio Neira thanks partial funding from COPAS Sur-Austral AFB 170006 and Núcleo Milenio INVASAL funded by Iniciativa Científica Milenio from Chile's Ministerio de Economía, Fomento y Turismo.

References

Alarcón, R., & Zuleta, A. (2013). Reporte de la evaluación de stock de merluza de cola (*Macruronus magellanicus*), 2012. *Centro de Estudios Pesqueros (CEPES), 2013*, 46.

Arancibia, H. (Ed.). (2015). *Hakes biology and exploitation*. London: Wiley. (ISBN: 9781118568415).

Arancibia, H., & Neira, S. (2005). Incorporación de indicadores cuantitativos en los objetivos sociales, económicos y ecológicos para la administración multi-específica de pesquerías en Chile central. In E. Figueroa (Ed.), *Biodiversidad marina: valoración, usos y perspectivas: Hacia dónde va Chile?* (pp. 539–553). Santiago, Chile: Editorial Universitaria. (ISBN: 956-11-1782-7).

Arancibia, H., & Neira, S. (2008). Overview of the Chilean hake *Merluccius gayi* stock, a biomass forecast, and the jumbo squid (*Dosidicus gigas*) predator–prey relationship off central Chile (33°S–39°S). *CalCOFI Rep, 49*, 104–115.

Arancibia, H., Neira, S., Barros, M., Gatica, C., Zúñiga, M. J., Alarcón, R., & Acuña, E. (2010). *Formulación e implementación de un enfoque multiespecífico de evaluación de stock en recursos demersales de la zona sur austral—Fase I. Informe Final Proyecto FIP 2008-23*. Universidad de Concepción/Instituto de Investigación Pesquera VIII Región S.A., p. 303.

Arancibia, H., Riquelme, J.A., Sagua, C., Barros, M., Alarcón, R., Neira, S., & Santis, O. (2016). Informe Final Corregido. Proyecto "Evaluación de los planes de recuperación de las pesquerías de merluza común y merluza del sur en un marco de modelación bioeconómico". Universidad de Concepción, p. 233.

Christensen, V., & Pauly, D. (1992). ECOPATH II. A software for balancing steady state ecosystem models and calculating network characteristics. *Ecol. Model., 61*, 169–185.

Christensen, V., Walters, C. J., & Pauly, D. (2005). *ECOPATH with ECOSIM: A user's guide* (p. 154). Vancouver, BC: Fisheries Centre, University of British Columbia.

Cochrane, K.L. (ed). (2002). *A fishery manager's guidebook. Management measures and their application* (*FAO Fisheries Technical Paper No. 424*, p. 231). Rome: FAO.

Cubillos, L., Alarcón, R., & Arancibia, H. (2007). Selectividad por tamaño de las presas en merluza común (*Merluccius gayi gayi*), zona centro-sur de Chile (1992–1997). *Invest. Mar. Valparaíso, 35*(1), 55–69.

FAO. (1995). *Code of conduct for responsible fisheries* (p. 41). Rome, Italy: FAO. ISBN 92-5-103834-5.

FAO. (1997). *Technical guidelines for responsible fisheries*. Rome, Italy: FAO. http://www.fao.org/docrep/003/w4230e/w4230e00.html.

FAO. (2003). Fisheries management 2. The ecosystem approach to fisheries. In *FAO technical guidelines for responsible fisheries* (p. 112). Rome: Food and Agricultural Organization of the United Nations.

FAO. (2008). Fisheries management. 2. The ecosystem approach to fisheries. 2.1 Best practices in ecosystem modelling for informing an ecosystem approach to fisheries. In *FAO fisheries technical guidelines for responsible fisheries* (Vol. 4, p. 78). Rome: Food and Agricultural Organization of the United Nations.

FAO. (2016). El estado mundial de la pesca y la acuicultura 2016. In *Contribución a la seguridad alimentaria y la nutrición para todos* (p. 224). Roma: FAO.

Gatica, C., Neira, S., Arancibia, H., & Vásquez, S. (2015). The biology, fishery and market of Chilean hake (Merluccius gayi gayi) in the southeastern Pacific Ocean. In H. Arancibia (Ed.), *Hakes biology and exploitation* (pp. 127–153). London: Wiley. (ISBN 9781118568415).

Hammer, C., von Dorrien, C., Hopkins, C. C. E., Köster, F. W., Nilssen, E. M., St John, M., & Wilson, D. C. (2010). Framework of stock-recovery strategies: Analyses of factors affecting success and failure. *ICES Journal of Marine Science, 67*, 1849–1855.

Husktadt, L. A., & Antezana, T. (2003). Behaviour of the southern sea lion (*Otaria flavescens*) and consumption of the catch during purse–seining for jack mackerel (*Trachurus symmetricus*) of central Chile. *ICES Journal of Marine Science, 60*, 1003–1011.

Hutchings, J. A., & Rangeley, R. W. (2011). Correlates of recovery for Canadian Atlantic cod (*Gadus morhua*). *Canadian Journal of Zoology, 89*, 386–400.

Ibañez, C. M., Arancibia, H., & Cubillos, L. A. (2008). Biases in determining the diet of jumbo squid *Dosidicus gigas* (D'Orbigny 1835) (Cephalopoda: Ommastrephidae) off southern-central Chile (34°S-40°S). *Helgoland Marine Research, 62*, 331–338.

Jurado-Molina, J., Gatica, C., Arancibia, H., Neira, S., & Alarcón, R. (2016). A multispecies virtual population analysis for the southern Chilean demersal fishery. *Marine and Coastal Fisheries, 8*(1), 350–360.

Jurado-Molina, J., Gatica, C., & Cubillos, L. (2006). Incorporating cannibalism into age-structured model for the Chilean hake. *Fishery Research, 82*, 30–40.

Neira, S., & Arancibia, H. (2004). Trophic interactions and community structure in the Central Chile marine ecosystem (33°S–39°S). *Journal of Experimental Marine Biology and Ecology, 312*, 349–366.

Neira, S., Arancibia, H., Alarcón, R, Barros, M., Cahuin, S., Castro, L., et al. (2012). *Informe Final Proyecto Subsecretaría de Pesca "Bases metodológicas para el estudio del reclutamiento y ecología en merluza del sur y merluza de cola, zona sur-austral* (ID 4728-23-LP11, p. 266 Anexos)". Universidad de Concepción.

Neira, S., Arancibia, H., & Cubillos, L. (2004). Comparative analysis of trophic structure of commercial fishery species off Central Chile in 1992 and 1998. *Ecological Modeling, 172*(2–4), 233–248.

Odum, E. P. (1969). The strategy of ecosystem development. *Science, 104*, 262–270.

Odum, E. P. (1971). *Fundamentals of Ecology* (p. 574). London: Saunders.

Odum, E. P. (1985). Trends expected in stressed ecosystems. *Bioscience, 35*(7), 419–422.

Quiroz, J. C., Ojeda, V., Chong, L., Lillo, S., & Céspedes, R. (2015). Estatus y posibilidades de explotación biológicamente sustentables de los principales recursos pesqueros nacionales

2016: Merluza del sur, 2016. Informe de Estatus (pp. 71). Instituto de Fomento Pesquero, IFOP, Valparaíso. Chile.

Payá, I. (1992). The diet of Patagonian hake *Merluccius australis polylepis* and its daily ratio of Patagonian grenadier *Macruronus magellanicus*. *South African Journal of Marine Science, 12*(1), 753–760.

Payá, I., Céspedes, R., Ojeda, V., Adasme, L., Gonzáles, J., & Lillo, S. (2014). *Informe estatus y cuota merluza de cola 2015* (p. 108). Valparaíso: Instituto de Fomento Pesquero.

Plagányi, É. E. (2007). *Models for an ecosystem approach to fisheries* (*FAO Technical Paper No. 477*, p. 108). Rome: FAO.

Schaefer, M. B. (1991). Some aspects of the dynamics of populations important to the management of the commercial marine fisheries. *Bulletin of Mathematical Biology, 53*(1–2), 253–279.

Tascheri, R., Cnales, C., Gálves, P., & Sateler. (2014). *Informe de estatus y cuota merluza común* (p. 116). Valparaíso: Instituto de Fomento Pesquero.

United Nations. (1995). United Nations Conference on Straddling Fish Stocks and Highly Migratory Fish Stocks : report of the Secretary-General. New York, 49 p.

United Nations. (1997). The Law of the Sea : official texts of the United Nations Convention on the Law of the Sea of 10 December 1982 and of the Agreement Relating to the Implementation of Part XI of the United Nations Convention on the Law of the Sea of 10 December 1982 with index and excerpts from the Final Act of the Third United Nations Conference on the Law of the Sea. New York, 294 p

United Nations. (2002). Report of the World Summit on Sustainable Development, Johannesburg, South Africa, 26 August-4 September 2002. New York, 170 p.

Urrutia, O., Parra, R., & Bermúdez, J. (2015). *Recuperación de pesquerías: Análisis de experiencias comparadas* (p. 153). Valparaíso, Chile: Ediciones Universitarias de Valparaíso, Pontificia Universidad Católica de Valparaíso.

Walters, C., Christensen, V., & Pauly, D. (1997). Structuring dynamic models of exploited ecosystems from trophic mass-balance assessments. *Reviews in Fish Biology and Fisheries, 7*, 139–172.

Macroscopic Properties and Keystone Species Complexes in Kelp Forest Ecosystems Along the North-Central Chilean Coast

Brenda B. Hermosillo-Núñez, Marco Ortiz, Ferenc Jordán, and Anett Endrédi

1 Introduction

Kelp forests in the subtidal rocky along the north-central Chilean coast have been heavily exploited, inducing drastic changes in the benthic communities at spatial and temporal scales (Ortiz & Levins, 2017; Vásquez et al., 2013). The kelp forest is mainly comprised of two species of Laminariales: *Macrocystis pyrifera* (Linnaeus) C. Agardh, 1820 and *Lessonia trabeculata* (Villouta & Santelices, 1986). Both species are considered ecosystem engineers (Jones, Lawton, & Shachak, 1994) and niche constructors (Laland, Odling-Smee, & Feldman, 1996; Levins & Lewontin, 1985), since their structure supplies areas for reproduction, food and refuge for many vertebrates and invertebrates (Steneck et al., 2002; Tegner & Dayton, 2000; Villegas, Laudien, Sielfeld, & Arntz, 2007). Although kelp forests provide impor-

B. B. Hermosillo-Núñez (✉)
Laboratorio de Modelamiento de Sistemas Ecológicos Complejos (LAMSEC), Instituto Antofagasta, Universidad de Antofagasta, Antofagasta, Chile
e-mail: Brenda.hermosillo@uantof.cl

M. Ortiz
Laboratorio de Modelamiento de Sistemas Ecológicos Complejos (LAMSEC), Instituto Antofagasta, Universidad de Antofagasta, Antofagasta, Chile

Instituto de Ciencias Naturales Alexander von Humboldt, Facultad de Ciencias del Mar y Recursos Biológicos, Universidad de Antofagasta, Antofagasta, Chile

Departamento de Biología Marina, Facultad de Ciencias del Mar, Universidad Católica del Norte, Coquimbo, Chile

F. Jordán
Balaton Limnological Institute, Centre for Ecological Research, Tihany, Hungary

Stazione Zoologica Anton Dohrn, Napoli, Italy

A. Endrédi
Balaton Limnological Institute, Centre for Ecological Research, Tihany, Hungary

M. Ortiz, F. Jordán (eds.), *Marine Coastal Ecosystems Modelling and Conservation*, https://doi.org/10.1007/978-3-030-58211-1_5

tant ecological and economic services to the human population, they are subject to many anthropogenic disturbances (Jackson et al., 2001; Lotze et al., 2006). Human exploitation has led to an increase in pollution, a degradation of ecosystems, and the deterioration of the species that inhabit these ecological systems (Halpern et al., 2008; Jackson et al., 2001). The north-central Chilean coast has various levels of disturbance triggered mainly by port activities, overfishing, tourism and urbanisation; these disturbances are associated with changes on the structure, organisation and performance of the ecosystem (González, Ortiz, Rodríguez-Zaragoza, & Ulanowicz, 2016; Pauly, Christensen, Dalsgaars, Froese, & Torres Jr, 1998; Petersen et al., 2008; Ray, Ulanowicz, Majee, & Roy, 2000).

1.1 Macroscopic or Emergent Ecosystem Properties

Systems theory helps to address novel kinds of questions about ecosystems that could not even be asked in terms of species (Odum, 1969; Odum & Odum, 1955). Trophic mass-balance models offer macroscopic descriptors for the dynamic and structure of ecosystems (Almunia, Basterretxea, Aristegui, & Ulanowicz, 1999; Costanza, 1992; Monaco & Ulanowicz, 1997; Ray et al., 2000), using the theoretical framework of Ulanowicz (1986, 1997) to assess the levels of development or maturity, organisation and health of ecosystems through *Ascendency*. Ecosystem development or maturity describes the maximum biomass (information content) and optimal energy utilisation of the ecosystem; organisation represents the number and diversity of interactions among components (Ulanowicz, 1986, 1997). A healthy ecosystem may be viewed as the one able to maintain its organisation and function over time (sensu Costanza & Mageau, 1999). These analyses have been widely used for assessing systemic properties of ecosystems in different geographical locations and at different levels of complexity. Hermosillo-Núñez, Ortiz, Rodríguez-Zaragoza, and Cupul-Magaña (2018) showed that areas of coral reef in the Mexican Pacific coast furthest from urbanisation present the best conditions of maturity, development and organisation based in theoretical framework of *Ascendency*. Likewise, González et al. (2016) used the theoretical frameworks of Odum (1969) and Ulanowicz (1986, 1997) to suggest that in the last 20 years, the benthic ecosystem of Tongoy Bay, Chile, has exhibited changes in macroscopic properties due to the reduction of human disturbances.

1.2 Keystone Species Complex: A Holistic Concept

Multispecies trophic network analyses allow the determination of species and/or functional groups that play key roles in ecological systems (Benerjee, Scharler, Fath, & Ray, 2017; Giacaman-Smith, Neira, & Arancibia, 2016; Jordán & Molnár, 1999; Jordán, Pereira, & Ortiz, 2019; Libralato, Christensen, & Pauly, 2006; Okey, 2004;

Ortiz et al., 2013; Ortiz et al., 2013; Ortiz et al., 2017; Valls, Coll, & Christensen, 2015). The ecological concept of keystone species (Paine, 1969) could be considered highly relevant information for designing biodiversity conservation monitoring and management programmes (Caro, 2010; Noss, 1999; Simberloff, 1998). However, their use in conservation management has generated controversy: keystone species have mainly been identified through manipulation experiments (Heske, Brown, & Mistry, 1994; Paine, 1969, 1974) or studies on the cascade effect (Estes & Palmisano, 1974; Estes, Tinker, Williams, & Doak, 1998) that have focused on few species and have not considered the complexity of the ecosystem (Jordán & Molnár, 1999; Jordán, Okey, Bauer, & Libralato, 2008; Ortiz et al., 2017; Ortiz, Levins, et al., 2013). In this work, an extended and holistic concept was used, based on the 'keystone species complex' (KSC) concept, described empirically (Daily, Ehrlich, & Haddad, 1993) and used in modelling efforts (Ortiz, Campos, et al., 2013). This holistic index quantitatively identifies a small set (core) of species (or functional groups) that are trophically related, being at different trophic levels, being relatively less abundant and influencing relatively strongly the rest of the species through direct and indirect chain effects (Fig. 1). The temporal extent of the KSC represents only transient dynamics, and it can exhibit changes through time. The KSC can assist management and conservation programmes with a focus of network trophic.

The aim of this chapter was to quantify keystone species complexes and macroscopic properties through trophic models that represent the ecological relationships

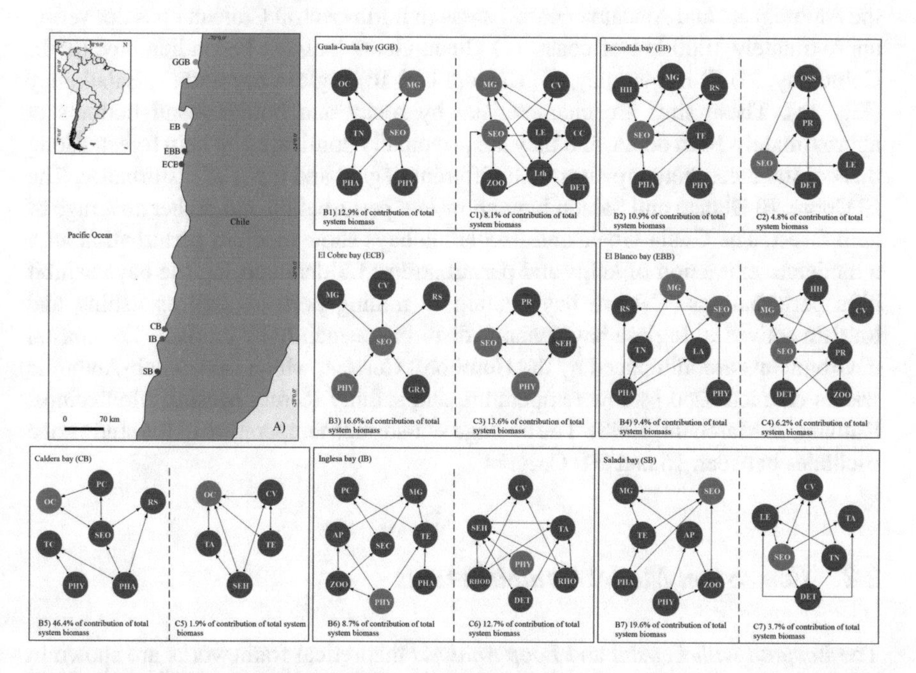

Fig. 1 (**a**) Study sites along the coast of northern Chile. (**b**) Keystone Species Complex obtained using quantitative and semi-quantitative indices. (**c**) Keystone Species Complex based on central node sets

(prey–predator/resource-consumer) among the most abundant species in seven kelp forest ecosystems of north-central Chilean coast with different human disturbance degrees. Based on this network analysis, the following ecological attributes were evaluated: (1) the emergent properties of the kelp forest ecosystems related to the structure, organisation and ecosystem health based on the theoretical frameworks of Ulanowicz (1986, 1997) and (2) the species and functional groups that play a key-stone role in the ecosystems and that should be conserved and/or managed for the well-functioning of the ecosystems. This investigation represents the first study based on mass-balanced trophic models for these ecological systems; the findings could improve the design of the current fishing management and conservation of diversity. It is important to note that all macroscopic ecosystem properties and KSCs describe a steady-state condition (balanced models) or represent only transient dynamics.

2 Geographical Areas Modelling Structure and Assumptions

2.1 Study Area

Field studies were conducted between 2007 and 2012 in seven small bays located in the Antofagasta and Atacama coastal areas in north-central Chilean coast, covering approximately 1000 km of coast: (1) Guala-Guala bay, (2) Escondida bay, (3) El Cobre bay, (4) El Blanco bay, (5) Caldera bay, (6) Inglesa bay and (7) Salado bay (Fig. 1a). These sites are characterised by rocky and boulder-sand bottoms at approximately 10 m depth, and they are principally dominated by kelp forest. Along this coastal area, there are sites with different degree and types of disturbance. The El Cobre, El Blanco and Salado bays show low perturbation and higher coverage of kelp forest. The Guala-Guala and Escondida bays show medium perturbation with a moderate extraction of kelps and perturbations. Caldera and Inglesa bays exhibit high perturbations. Caldera bay contains a mining port, as well as fishing and tourism activities; Inglesa bay contains disturbance mainly by tourism. The coastal environments are influenced by the Humboldt Current, which carries sub-Antarctic waters characterised by low temperature and salinity (Strub, Mesías, Montecinos, Rutllant, & Marchant, 1998). The average sea-surface temperature in the study zone oscillates between 16 and 20 °C.

2.2 Ecosystem Model Compartments

The *Ecopath with Ecosim* and *Loop Analysis* theoretical frameworks are shown in Boxes 1 and 2. The constructed models considered the trophic relationships of the most relevant species and functional groups inhabiting the benthic communities.

The seven models were composed of 21–26 compartments representing individual species that were characterised by high density and fishing importance or were representative species of the ecological system; for example, the starfish *Meyenaster gelatinosus* (Meyen, 1834), and *Heliaster helianthus* (Lamarck, 1816), *Octopus* sp. Cuvier, 1797, the fish *Cheilodactylus variegatus* Valenciennes, 1833, and *Pinguipes chilensis* Valenciennes, 1833, the crab *Romaleon setosum,* (Poeppig, 1836) the gastropods *Priene scabrum,* (P. P. King, 1832), *Concholepas concholepas* (Bruguière, 1789), *Turritella cingulata* G. B. Sowerby I, 1825 and *Tegula* sp., as well as the urchins *Loxechinus albus* (Molina, 1782) and *Tetrapygus niger* (Molina, 1782), the tunicate *Pyura chilensis* Molina, 1782, the mollusc *Leukoma thaca* (Molina, 1782), and the kelp species *Lessonia trabeculata.* Likewise, functional groups included several species that were selected for food preferences and trophic relationships, for example, the functional group Other starfish (OS), Small epifauna carnivores (SEC), Small epifauna herbivores (SEH), Small epifauna omnivores (SEO), Large epifauna (LE), Phaeophyta (PHA), Rhodophyta (RHO), and Chlorophyta (CHL). The species varied in the sites of study along the northern Chilean coast (Table 1).

Box 1 Ecopath with Ecosim Theoretical Framework

The energy mass balance of a species or functional group within a network is represented by the following mathematical expression:

$$B_i\left(\frac{P}{B}\right)_i EE_i - \sum_{j=1}^{n} B_j\left(\frac{Q}{B}\right)_j DC_{ji} - Y_i - BA_i - E_i = 0, \qquad (1)$$

where B_i and B_j are the biomass of prey i and predator j; P/B_i is the productivity of prey i, which is equivalent to total mortality (Allen, 1971); EE_i is the ecotrophic efficiency or the fraction of the total production of a group or species used in the system; Y_i is the fishing production per unit of area and time (Y = fishing mortality x biomass); Q/B_i is the food consumption per biomass unit j; DC_{ji} is the fraction of prey i in the average diet of predator j; BA_i is the biomass accumulation rate for i; and E_i is the net migration of i (emigration minus immigration) (Christensen and Walters, 2004).

Based on this equation, the input and output of matter (energy) in each compartment of the system can be balanced. This energy balance is assured for each group with this equation:

$$Q = P + R + UAF, \qquad (2)$$

where Q is the consumption, P is the production, R is the respiration and UAF is the unassimilated food for each group or species of the system. Given the inclusion of the factors BA_i and E_i in Eq. (1), the focus of *Ecopath* is based more on energetic continuity than a stable state condition. This situation

Table 1 Species composition of functional groups in the sites of study along the northern Chilean coast

Functional group	Species
Other starfish (OS)	*Patiria chilensis* (Lutken, 1859)
	Luidia magellanica (Leipoldt, 1895)
	Stichaster striatus (Müller & Troschel, 1840)
Small epifauna carnivores (SEC)	*Crassilabrum crassilabrum* (G. B. Sowerby II, 1834)
	Alia unifasciata (G. B. Sowerby I, 1832)
	Nassarius gayii (Kiener, 1834)
	Felicioliva peruviana (Lamarck, 1811)
	Xanthochorus sp. (Fischer, 1884)
	Allopetrolisthes punctatus (Guérin, 1835)
	Nudibranch
	Polynoidae
	Eunicidae
Small epifauna herbivores (SEH)	*Chaetopleura benaventei* (Plate, 1899)
	Tonicia elegans (Frembly, 1827)
	Chiton cumingi (Frembly, 1827)
	Fissurella crassa (Lamarck, 1822)
	Fissurella latimarginata (Sowerby I, 1835)
Small epifauna omnivores (SEO)	*Blepharipoda spinosa* (Milne Edwards & Lucas, 1841)
	Emerita analoga (Stimpson, 1857)
	Platymera gaudichaudii (Milne Edwards, 1837)
	Caprella sp. (Lamarck, 1801)
	Anemonia alicemartinae (Häussermann & Försterra, 2001)
	Spionidae sp. (Grube 1850)
	Polinices uber (Valenciennes, 1832)
	Rissoina inca (d'Orbigny, 1841)
	Spisula solida (Linnaeus, 1758)
	Balanus laevis (Bruguiére, 1789)
	Nereidae
	Cnidaria
	Amphipoda
	Bryozoa
Large epifauna (LE)	*Pagurus edwardsii* (Dana, 1852)
	Pagurus villosus (Nicolet, 1849)
	Tetralobistes weddellii (H. Milne Edwards, 1848)
	Paraxanthus barbiger (Poeppig, 1836)
	Acanthonyx petiveri (Milne Edwards, 1834)
	Pseudocorystes sicarius (Poeppig, 1836)
	Petrolisthes desmarestii (Guérin, 1835)
	Ophiuroidea
	Holothuroidea

(continued)

Table 1 (continued)

Functional group	Species
Phaeophyta (PHA)	*Dyctiota* sp. (Lamouroux, 1809)
	Desmarestia ligulata (Lamouroux, 1813)
	Halopteris sp. (Allman, 1877)
Rhodophyta (RHO)	*Chondrus* sp. (Stackhouse, 1797)
	Gigartina sp. (Starckhouse, 1809)
	Gracilaria sp. (Greville, 1830)
	Polysiphonia sp. (Greville, 1823)
Chlorophyta (CHL)	*Ulva* sp. (Linnaeus, 1753)

allows changes in the network compartments when the mathematical expression is expressed in a dynamic form.

To employ *Ecosim* (see Walters and Christensen, 2007), an extension routine of *Ecopath* is included to define the consumption by compartment (Q_{ij}), where Q_{ij} is represented by the following equation:

$$Q_{ij} = \frac{a_{ij} v_{ij} B_i B_j}{2v_{ij} + a_{ij} B_j} \quad (3)$$

where a_{ij} represents the instantaneous mortality rate on prey *i* caused by a single unit of predator *j* biomass. Similarly, a_{ij} can be understood as the rate of effective search by predator *j* for prey *i*. Each a_{ij} is estimated directly from the corresponding *Ecopath* models by $a_{ij} = Q_i/(B_i B_j)$, where Q_i is the total consumption of *i*. The v_{ij} represents the transference rate between compartment *i* and *j*. This parameter determines if the flow control mechanism is top-down, bottom-up or mixed.

Box 2 Loop Analysis Theoretical Framework

Loop Analysis estimates the local stability (as a measure of sustainability) of an ecological system and the assessment of the propagation of both direct and indirect effects as a response to external perturbations (Levins, 1974, 1998). The interactions are shown as signs that indicate the type of influence each variable has upon another (positive, negative, or null). In ecological relationships, the +/− signs denote predator/prey or parasite/host interactions, +/+ signs express mutualism, +/0 signs represent commensalism and −/0 signs show amensalism. *Loop Analysis* is based on the relationships of differential equations near equilibrium, community matrices and their loop diagrams. In a benthic system, the element αij of the matrix and loop diagrams performs the effect of the variable j in the growing of variable i, and this dynamic performs in the following way:

$$\frac{dX_i}{dt} = f_i\left(X_1, X_2, \ldots, X_n; C_1, C_2, \ldots, C_n\right), \quad (4)$$

where the change on the time of variable Xi is a function fi of the other variables—Xn and parameters Cn—which are interconnected. The link of Xj to Xi is similarly to $\propto ij$ in Levins (1968):

$$\propto ij = \frac{\partial fi\left(X\right)X*}{\partial Xj}, \quad (5)$$

where x^* indicates that it has been evaluated in equilibrium. The sign $\propto ij$ represents the link of j to i where the function of sign X is 1 when $X > 0$, 0 when $X = 0$ and -1 when $X < 0$.

Local stability of the system is quantified using the Routh–Hurwitz criteria, which require the following: (1) all negative feedbacks and (2) negative feedbacks on higher levels cannot be too great for comparison with the negative feedbacks on lower levels. Levins (1998) proposes that the system is more resistant (locally stable), whereas Fn (feedback corresponding to higher level of complexity) is more negative.

$$F_0\lambda^n + F_1\lambda^{n-1} + F_2\lambda^{n-2} \ldots. + F_{n-1}\lambda + F_n = 0. \quad (6)$$

Seasonal samplings were performed to estimate the average biomass (B) and density of macrobenthos and fish species (±10 m depth). Production (P) and turnover rates (P/B) were estimated following the allometric Eq. (7):

$$\text{Production} = \left[\left(\frac{\text{Biomass}}{\text{Density}}\right)^{0.73}\right] \times \text{Density}, \quad (7)$$

where 0.73 is the average regression exponent of annual production on body-size for macrobenthic invertebrates (sensu Warwick & Clarke, 1993).

Estimations of turnover rates (P/B) were obtained 10 length frequencies, which were used to estimate the initial growth parameters (K and L∞) and adjusted to the von Bertalanffy growth function. Once the growth parameters were calculated, the instantaneous rate of total mortality (Z) was estimated using the length-converted catch equation (Sparre & Venema, 1997). It is important to note that Z is used to approximate P/B (after Allen, 1971). For some organisms, the consumption rate (Q/B) was calculated for a 24-h cycle following the procedure described by Cerda and Wolff (1993); for other organisms, the Q/B and capture estimates were taken

Table 2 Parameter values entered (in bold) and estimated (standard) by *Ecopath* software for the benthic-pelagic coastal ecological systems in Atacama and Antofagasta Regions of Chile. *TL* level trophic, *Ca* capture, *B* biomass (*g* wet weight [ww] m^{-2}), P/B = turnover rate ($year^{-1}$), Q/B = consumption rate ($year^{-1}$), *EE* ecotrophic efficiency (dimensionless), *GE* gross efficiency (dimensionless), *NE* net efficiency (dimensionless), *RA/AS respiration/assimilation rate* (dimensionless), *RA/B* respiration/biomass rate ($year^{-1}$), *P/RA* production/respiration rate (dimensionless)

	Group/specie name	TL	Ca	B	P/B	Q/B	EE	GE	NE	RA/AS	RA/B	P/RA
Guala-Guala bay												
1)	*Meyenaster gelatinosus* (MG)	3.18		**14.00**	**1.20**	**5.00**	0.24	0.24	0.30	0.70	2.80	0.43
2)	*Heliaster helianthus* (HH)	3.16		**35.50**	**1.40**	**4.80**	0.06	0.29	0.36	0.64	2.44	0.57
3)	Other starfish (OS)	3.21		**4.50**	**1.50**	**5.00**	0.39	0.30	0.38	0.63	2.50	0.60
4)	*Octopus* sp (OC)	3.74	**21.20**	**29.50**	**2.50**	**9.00**	0.99	0.28	0.35	0.65	4.70	0.53
5)	*Cheilodactylus variegatus* (CV)	3.27	**0.10**	**16.86**	**2.10**	**7.50**	0.98	0.28	0.35	0.65	3.90	0.54
6)	*Pinguipes chilensis* (PC)	3.29	**0.10**	**17.50**	**2.10**	**7.50**	0.95	0.28	0.35	0.65	3.90	0.54
7)	*Taliepus* sp (TA)	2.15		**17.50**	**2.80**	**9.50**	0.97	0.29	0.37	0.63	4.80	0.58
8)	*Concholepas concholepas* (CC)	3.39	**0.33**	**32.50**	**3.20**	**11.00**	0.97	0.29	0.36	0.64	5.60	0.57
9)	*Priene* sp (PR)	2.81		**50.41**	**2.00**	**10.40**	0.05	0.19	0.24	0.76	6.32	0.32
10)	Large epifauna (LE)	2.89		**75.50**	**2.50**	**9.50**	0.91	0.26	0.33	0.67	5.10	0.49
11)	Small epifauna carnivore (SEC)	2.97		**140.50**	**3.70**	**12.50**	0.98	0.30	0.37	0.63	6.30	0.59
12)	Small epifauna omnivore (SEO)	2.40		**419.50**	**3.90**	**14.50**	0.98	0.27	0.34	0.66	7.70	0.51
13)	Small epifauna herbivore (SEH)	2.00		**120.50**	**3.60**	**12.50**	0.99	0.29	0.36	0.64	6.40	0.56
14)	*Turritella cingulata* (TC)	2.00		**150.50**	**4.10**	**14.50**	0.35	0.28	0.35	0.65	7.50	0.55
15)	*Tegula* sp (TE)	2.00		**290.50**	**4.30**	**14.50**	0.18	0.30	0.37	0.63	7.30	0.59
16)	*Tetrapygus niger* (TN)	2.00		**83.50**	**3.10**	**10.50**	0.96	0.30	0.37	0.63	5.30	0.58
17)	*Loxechinus albus* (LA)	2.00	**0.58**	**14.50**	**2.80**	**9.50**	0.99	0.29	0.37	0.63	4.80	0.58
18)	*Protothaca thaca* (PT)	2.10	**15.70**	**55.50**	**3.30**	**11.00**	0.93	0.30	0.38	0.63	5.50	0.60
19)	*Pyura chilensis* (PyC)	2.40	**0.29**	**252.70**	**3.20**	**11.00**	0.18	0.29	0.36	0.64	5.60	0.57
20)	*Lessonia trabeculata* (LT)	1.00	**94.80**	**5512.50**	**3.40**		0.13					
21)	Phaeophyta (PHA)	1.00		**435.50**	**8.50**		0.98					

(continued)

Table 2 (continued)

	Group/specie name	TL	Ca	B	P/B	Q/B	EE	GE	NE	RA/AS	RA/B	P/RA
22)	Zooplankton (ZOO)	2.00		**35.00**	**80.00**	**270.00**	0.87	0.30	0.37	0.63	136.00	0.59
23)	Phytoplankton (PHY)	1.00		**50.00**	**280.00**		0.94					
24)	Detritus (DET)	1.00		**100.00**			0.26					
Escondida bay												
1)	*Meyenaster gelatinosus* (MG)	3.4		**8.0**	**0.6**	**5.0**	0.79	0.12	0.15	0.85	3.40	0.18
2)	*Heliaster helianthus* (HH)	3.4		**35.5**	**1.4**	**4.8**	0.05	0.29	0.36	0.64	2.44	0.57
3)	Other starfish (OS)	3.3		**3.5**	**1.5**	**5.0**	0.43	0.30	0.38	0.63	2.50	0.60
4)	*Cheilodactylus variegatus* (CV)	3.3	**1.38**	**8.9**	**2.1**	**7.5**	0.18	0.27	0.34	0.66	3.95	0.52
5)	*Pinguipes chilensis* (PC)	3.4	**1**	**8.5**	**2.1**	**7.5**	0.13	0.28	0.35	0.65	3.90	0.54
6)	*Romaleon polyodon* (RP)	3.2	**10.4**	**30.1**	**1.1**	**9.5**	1.00	0.12	0.14	0.86	6.50	0.17
7)	*Concholepas concholepas* (CC)	3.4	**0.1**	**21.5**	**3.2**	**11.0**	0.95	0.29	0.36	0.64	5.60	0.57
8)	*Priene* sp (PR)	2.9		**11.7**	**2.0**	**10.4**	0.15	0.19	0.24	0.76	6.32	0.32
9)	Large epifauna (LE)	2.9		**45.5**	**2.5**	**9.5**	0.90	0.26	0.33	0.67	5.10	0.49
10)	Small epifauna carnivore (SEC)	3.0		**108.5**	**3.7**	**12.5**	0.99	0.30	0.37	0.63	6.30	0.59
11)	Small epifauna omnivore (SEO)	2.4		**295.5**	**3.9**	**14.5**	1.00	0.27	0.34	0.66	7.70	0.51
12)	Small epifauna herbivore (SEH)	2.2		**95.5**	**3.6**	**12.5**	0.98	0.29	0.36	0.64	6.40	0.56
13)	*Turritella cingulata* (TC)	2.0		**150.5**	**3.9**	**14.5**	0.36	0.27	0.34	0.66	7.70	0.51
14)	*Tegula* sp (TE)	2.0		**350.0**	**1.3**	**14.5**	0.51	0.09	0.11	0.89	10.33	0.12
15)	*Loxechinus albus* (LA)	2.0	**50.7**	**47.5**	**3.1**	**10.4**	0.99	0.30	0.37	0.63	5.22	0.59
16)	*Leukoma thaca (LT)*	2.1	**10.8**	**30.5**	**3.3**	**11.0**	0.99	0.30	0.38	0.63	5.50	0.60
17)	*Gari solida* (GS)	2.1		**6.6**	**1.2**	**7.0**	0.69	0.17	0.21	0.79	4.40	0.27
18)	*Pyura chilensis* (PyC)	2.4	**4.4**	**350.5**	**3.2**	**11.0**	0.10	0.29	0.36	0.64	5.60	0.57
19)	*Lessonia trabeculata* (LT)	1.0	**156**	**7408.3**	**3.4**	**0.0**	0.14					
20)	Phaeophyta (PHA)	1.0		**270.0**	**5.0**	**0.0**	0.99					
21)	Zooplankton (ZOO)	2.0		**35.0**	**80.0**	**270.0**	0.95	0.30	0.37	0.63	136.00	0.59

	Group/specie name	TL	Ca	B	P/B	Q/B	EE	GE	NE	RA/AS	RA/B	P/RA
22)	Phytoplankton (PHY)	1.0		**50.0**	**280.0**	**0.0**	0.97					
23)	Detritus (DET)	1.0		**100.0**			0.21					
El Cobre bay												
1)	*Meyenaster gelatinosus* (MG)	3.3		**6.3**	**0.6**	**5.0**	0.08	0.12	0.15	0.85	3.40	0.18
2)	Other starfish (OS)	3.4		**2.1**	**1.5**	**5.0**	0.23	0.30	0.38	0.63	2.50	0.60
3)	*Cheilodactylus variegatus* (CV)	3.2	**0.924**	**8.9**	**2.1**	**7.5**	0.05	0.28	0.35	0.65	3.90	0.54
4)	*Pinguipes chilensis* (PC)	3.3	**0.192**	**3.1**	**2.1**	**7.5**	0.03	0.28	0.35	0.65	3.90	0.54
5)	*Romaleon polyodon* (RP)	3.1	**9.05**	**15.5**	**1.1**	**9.5**	0.62	0.12	0.14	0.86	6.50	0.17
6)	*Taliepus* sp (TA)	2.0		**11.7**	**2.8**	**9.5**	0.32	0.29	0.37	0.63	4.80	0.58
7)	*Concholepas concholepas* (CC)	3.7	**0.019**	**23.9**	**3.2**	**11.0**	0.99	0.29	0.36	0.64	5.60	0.57
8)	*Priene* sp (PR)	2.8		**11.2**	**2.0**	**10.4**	0.16	0.19	0.24	0.76	6.32	0.32
9)	Large epifauna (LE)	2.9		**24.4**	**2.5**	**9.5**	0.80	0.26	0.33	0.67	5.10	0.49
10)	Small epifauna carnivore (SEC)	2.9		**145.5**	**3.9**	**14.5**	0.99	0.27	0.34	0.66	7.70	0.51
11)	Small epifauna omnivore (SEO)	2.3		**340.5**	**3.9**	**14.5**	0.99	0.27	0.34	0.66	7.70	0.51
12)	Small epifauna herbivore (SEH)	2.0		**110.5**	**3.6**	**12.5**	0.91	0.29	0.36	0.64	6.40	0.56
13)	*Turritella cingulata* (TC)	2.0		**285.8**	**1.0**	**14.5**	0.57	0.07	0.09	0.91	10.60	0.09
14)	*Tegula* sp (TE)	2.0		**80.5**	**4.3**	**14.5**	0.47	0.30	0.37	0.63	7.30	0.59
15)	*Leukoma thaca (LT)*	2.1	**10.02**	**29.5**	**3.3**	**11.0**	0.90	0.30	0.38	0.63	5.50	0.60
16)	*Lessonia trabeculata* (LT)	1.0	**105.6**	**2000.7**	**3.4**	**0.0**	0.53					
17)	*Gracilaria* sp (GR)	1.0	**756.1**	**199.4**	**8.5**	**0.0**	0.98					
18)	*Rhodhymenia* sp (RHO)	1.0		**148.6**	**8.5**	**0.0**	0.53					
19)	Zooplankton (ZOO)	2.0		**18.0**	**80.0**	**270.0**	0.44	0.30	0.37	0.63	136.00	0.59
20)	Phytoplankton (PHY)	1.0		**28.0**	**280.0**	**0.0**	0.81					
21)	Detritus (DET)	1.0		**100.0**			0.48					
El Blanco bay												

(continued)

Table 2 (continued)

	Group/specie name	TL	Ca	B	P/B	Q/B	EE	GE	NE	RA/AS	RA/B	P/RA
1)	*Meyenaster gelatinosus* (MG)	3.2		**3.8**	**1.2**	**5.0**	0.08	0.24	0.30	0.70	2.80	0.43
2)	*Heliaster helianthus* (HH)	3.2		**2.1**	**1.4**	**4.8**	0.16	0.29	0.36	0.64	2.44	0.57
3)	Other starfish (OS)	3.2		**6.9**	**1.5**	**5.0**	0.06	0.30	0.38	0.63	2.50	0.60
4)	*Cheilodactylus variegatus* (CV)	3.2	**0.17**	**17.7**	**2.1**	**7.5**	0.02	0.28	0.35	0.65	3.90	0.54
5)	*Pinguipes chilensis* (PC)	3.4	**0.1**	**3.1**	**2.1**	**7.5**	0.05	0.28	0.35	0.65	3.90	0.54
6)	*Romaleon polyodon* (RP)	3.1	**1.6**	**18.8**	**0.4**	**9.5**	0.97	0.04	0.05	0.95	7.21	0.05
7)	*Taliepus* sp (TA)	2.0		**4.9**	**2.8**	**9.5**	0.93	0.29	0.37	0.63	4.80	0.58
8)	*Concholepas concholepas* (CC)	3.4	**0.01**	**12.5**	**3.2**	**11.0**	0.96	0.29	0.36	0.64	5.60	0.57
9)	*Priene* sp (PR)	2.8		**24.0**	**2.0**	**10.4**	0.11	0.19	0.24	0.76	6.32	0.32
10)	Large epifauna (LE)	2.9		**34.8**	**2.5**	**9.5**	0.64	0.26	0.33	0.67	5.10	0.49
11)	Small epifauna carnivore (SEC)	3.0		**65.3**	**3.7**	**12.5**	0.98	0.30	0.37	0.63	6.30	0.59
12)	Small epifauna omnivore (SEO)	2.4		**199.6**	**3.9**	**14.5**	0.99	0.27	0.34	0.66	7.70	0.51
13)	Small epifauna herbivore (SEH)	2.0		**229.2**	**3.6**	**12.5**	0.29	0.29	0.36	0.64	6.40	0.56
14)	*Turritella cingulata* (TC)	2.0		**493.1**	**4.1**	**14.5**	0.06	0.28	0.35	0.65	7.50	0.55
15)	*Tegula* sp (TE)	2.0		**40.9**	**4.3**	**14.5**	0.69	0.30	0.37	0.63	7.30	0.59
16)	*Tetrapygus niger* (TN)	2.0		**32.0**	**3.1**	**10.5**	0.94	0.30	0.37	0.63	5.30	0.58
17)	*Loxechinus albus* (LA)	2.0	**6.9**	**12.9**	**2.8**	**9.5**	0.89	0.29	0.37	0.63	4.80	0.58
18)	*Leukoma thaca (LT)*	2.1	**1.8**	**15.1**	**3.3**	**11.0**	0.96	0.30	0.38	0.63	5.50	0.60
19)	*Pyura chilensis* (PyC)	2.4	**0.5**	**32.2**	**3.2**	**11.0**	0.58	0.29	0.36	0.64	5.60	0.57
20)	*Lessonia trabeculata* (LT)	1.0	**19.4**	**4011.1**	**3.4**	**0.0**	0.31					
21)	Rhodophyta (RH)	1.0		**152.6**	**8.5**	**0.0**	0.99					
22)	Phaeophyta (PHA)	1.0		**254.1**	**8.5**	**0.0**	1.00					
23)	Zooplankton (ZOO)	2.0		**18.0**	**80.0**	**270.0**	0.56	0.30	0.37	0.63	136.00	0.59
24)	Phytoplankton (PHY)	1.0		**28.0**	**280.0**	**0.0**	0.78					
25)	Detritus (DET)	1.0		**100.0**			0.26					

	Group/specie name	TL	Ca	B	P/B	Q/B	EE	GE	NE	RA/AS	RA/B	P/RA
Caldera bay												
1)	*Meyenaster gelatinosus* (MG)	3.2		**11.2**	**1.2**	**5.0**	0.16	0.24	0.30	0.70	2.80	0.43
2)	*Heliaster helianthus* (HH)	3.4		**3.8**	**1.4**	**4.8**	0.25	0.29	0.36	0.64	2.44	0.57
3)	Other starfish (OS)	3.3		**1.6**	**1.5**	**5.0**	0.35	0.30	0.38	0.63	2.50	0.60
4)	*Octopus* sp (OC)	3.8	**1.30**	**0.6**	**2.5**	**9.0**	0.67	0.28	0.35	0.65	4.70	0.53
5)	*Cheilodactylus variegatus* (CV)	3.2	**0.33**	**1.9**	**2.1**	**7.5**	0.23	0.28	0.35	0.65	3.90	0.54
6)	*Pinguipes chilensis* (PC)	3.3	**0.66**	**1.1**	**2.1**	**7.5**	0.48	0.28	0.35	0.65	3.90	0.54
7)	*Romaleon polyodon* (RP)	3.1	**5.00**	**5.5**	**1.1**	**9.5**	0.66	0.12	0.14	0.86	6.50	0.17
8)	*Taliepus* sp (TA)	2.0		**3.6**	**2.8**	**9.5**	0.52	0.29	0.37	0.63	4.80	0.58
9)	*Concholepas concholepas* (CC)	3.7	**14.00**	**198.4**	**3.2**	**11.0**	0.99	0.29	0.36	0.64	5.60	0.57
10)	*Priene* sp (PR)	2.7		**710.0**	**2.0**	**10.4**	0.00	0.19	0.24	0.76	6.32	0.32
11)	Large epifauna (LE)	2.9		**49.1**	**2.5**	**9.5**	0.98	0.26	0.33	0.67	5.10	0.49
12)	Small epifauna carnivore (SEC)	3.0		**1095.0**	**4.7**	**16.5**	0.98	0.28	0.36	0.64	8.50	0.55
13)	Small epifauna omnivore (SEO)	2.4		**2475.0**	**4.9**	**16.5**	0.97	0.30	0.37	0.63	8.30	0.59
14)	Small epifauna herbivore (SEH)	2.0		**660.0**	**4.6**	**15.5**	0.99	0.30	0.37	0.63	7.80	0.59
15)	*Turritella cingulata* (TC)	2.0		**19206.9**	**4.1**	**14.5**	0.01	0.28	0.35	0.65	7.50	0.55
16)	*Tegula* sp (TE)	2.0		**588.9**	**4.3**	**14.5**	0.36	0.30	0.37	0.63	7.30	0.59
17)	*Argopecten purpuratus* (AP)	2.0	**0.50**	**65.1**	**3.3**	**11.0**	0.97	0.30	0.38	0.63	5.50	0.60
18)	*Rhodymenia* sp (RHO)	1.0		**38.7**	**5.0**	**0.0**	0.45					
19)	*Lessonia trabeculata* (LT)	1.0	**0.50**	**22700.0**	**4.4**	**0.0**	1.00					
20)	*Ulva* sp. (UL)	1.0		**2090.0**	**6.0**	**0.0**	0.40					
21)	Rhodophyta (RH)	1.0		**5990.0**	**6.5**	**0.0**	0.82					
22)	Phaeophyta (PHA)	1.0		**7992.0**	**8.5**	**0.0**	0.84					
23)	Zooplankton (ZOO)	2.0		**105.0**	**80.0**	**270.0**	0.99	0.30	0.37	0.63	136.00	0.59
24)	Phytoplankton (PHY)	1.0		**205.0**	**280.0**	**0.0**	0.93					

(continued)

Table 2 (continued)

	Group/specie name	TL	Ca	B	P/B	Q/B	EE	GE	NE	RA/AS	RA/B	P/RA
25)	Detritus (DET)	1.0		**100.0**			0.62					
Inglesa bay												
1)	*Meyenaster gelatinosus* (MG)	3.2		**186.9**	**1.2**	**5.0**	0.09	0.24	0.30	0.70	2.80	0.43
2)	*Heliaster helianthus* (HH)	3.4		**69.4**	**1.4**	**4.8**	0.23	0.29	0.36	0.64	2.44	0.57
3)	Other starfish (OS)	3.3		**20.3**	**1.5**	**5.0**	0.45	0.30	0.38	0.63	2.50	0.60
4)	*Cheilodactylus variegatus* (CV)	3.2	**0.30**	**8.9**	**2.1**	**7.5**	0.00	0.28	0.35	0.65	3.90	0.54
5)	*Pinguipes chilensis* (PC)	3.3	**0.60**	**3.1**	**2.1**	**7.5**	0.03	0.28	0.35	0.65	3.90	0.54
6)	*Romaleon polyodon* (RP)	3.2	**4.50**	**12.0**	**1.1**	**9.5**	0.19	0.12	0.14	0.86	6.50	0.17
7)	*Taliepus* sp (TA)	2.1		**8.3**	**2.8**	**9.5**	0.89	0.29	0.37	0.63	4.80	0.58
8)	*Priene* sp (PR)	2.7		**23.0**	**2.0**	**10.4**	0.30	0.19	0.24	0.76	6.32	0.32
9)	Large epifauna (LE)	2.9		**94.1**	**2.5**	**9.5**	0.95	0.26	0.33	0.67	5.10	0.49
10)	Small epifauna carnivore (SEC)	3.0		**152.9**	**3.7**	**12.5**	0.99	0.30	0.37	0.63	6.30	0.59
11)	Small epifauna omnivore (SEO)	2.4		**375.1**	**3.9**	**14.5**	1.00	0.27	0.34	0.66	7.70	0.51
12)	Small epifauna herbivore (SEH)	2.0		**138.3**	**3.6**	**12.5**	0.96	0.29	0.36	0.64	6.40	0.56
13)	*Turritella cingulata* (TC)	2.0		**590.9**	**4.1**	**14.5**	0.19	0.28	0.35	0.65	7.50	0.55
14)	*Tegula* sp (TE)	2.0		**110.8**	**4.3**	**14.5**	0.99	0.30	0.37	0.63	7.30	0.59
15)	*Austromegabalanus psittacus* (AP)	2.1	**0.01**	**49.6**	**1.1**	**7.0**	0.51	0.16	0.20	0.80	4.50	0.24
16)	*Argopecten purpuratus* (AP)	2.0	**0.45**	**91.5**	**3.3**	**11.0**	0.98	0.30	0.38	0.63	5.50	0.60
17)	*Tagelus dombeii* (TD)	2.0	**0.01**	**25.3**	**1.3**	**9.9**	0.76	0.13	0.16	0.84	6.62	0.20
18)	*Lessonia trabeculata* (LT)	1.0	**0.45**	**197.2**	**3.4**	**0.0**	0.97					
19)	*Ulva* sp. (UL)	1.0		**4173.8**	**5.0**	**0.0**	0.02					
20)	*Rhodymenia* sp. (RHO)	1.0		**663.3**	**5.0**	**0.0**	0.06					
21)	Rhodophyta (RH)	1.0		**669.1**	**5.5**	**0.0**	0.26					
22)	*Gracilaria* sp. (GR)	1.0	**0.43**	**4313.5**	**5.0**	**0.0**	0.04					
23)	Phaeophyta (PHA)	1.0		**550.7**	**8.5**	**0.0**	0.93					

	Group/specie name	TL	Ca	B	P/B	Q/B	EE	GE	NE	RA/AS	RA/B	P/RA
24)	Zooplankton (ZOO)	2.0		**18.9**	**80.0**	**270.0**	0.98	0.30	0.37	0.63	136.00	0.59
25)	Phytoplankton (PHY)	1.0		**30.8**	**280.0**	**0.0**	0.97					
26)	Detritus (DET)	1.0		**100.0**			0.13					
Salado bay												
1)	*Meyenaster gelatinosus* (MG)	3.2		**6.3**	**1.2**	**5.0**	0.20	0.24	0.30	0.70	2.80	0.43
2)	*Heliaster helianthus* (HH)	3.2		**12.6**	**1.4**	**4.8**	0.07	0.29	0.36	0.64	2.44	0.57
3)	Other starfish (OS)	3.2		**9.5**	**1.5**	**5.0**	0.06	0.30	0.38	0.63	2.50	0.60
4)	*Cheilodactylus variegatus* (CV)	3.2	**0.04**	**8.9**	**2.1**	**7.5**	0.00	0.28	0.35	0.65	3.90	0.54
5)	*Pinguipes chilensis* (PC)	3.3	**0.08**	**3.1**	**2.1**	**7.5**	0.01	0.28	0.35	0.65	3.90	0.54
6)	*Romaleon polyodon* (RP)	3.2	**0.62**	**2.9**	**1.1**	**9.5**	0.22	0.12	0.14	0.86	6.50	0.17
7)	*Taliepus* sp (TA)	2.0		**1.7**	**2.8**	**9.5**	0.58	0.29	0.37	0.63	4.80	0.58
8)	*Priene* sp (PR)	2.7		**41.3**	**2.0**	**10.4**	0.02	0.19	0.24	0.76	6.32	0.32
9)	Large epifauna (LE)	2.9		**26.5**	**2.5**	**9.5**	0.61	0.26	0.33	0.67	5.10	0.49
10)	Small epifauna carnivore (SEC)	3.0		**106.0**	**3.7**	**12.5**	0.77	0.30	0.37	0.63	6.30	0.59
11)	Small epifauna omnivore (SEO)	2.4		**252.4**	**3.9**	**14.5**	0.92	0.27	0.34	0.66	7.70	0.51
12)	Small epifauna herbivore (SEH)	2.0		**73.7**	**3.6**	**12.5**	0.95	0.29	0.36	0.64	6.40	0.56
13)	*Turritella cingulata* (TC)	2.0		**3970.3**	**4.1**	**14.5**	0.01	0.28	0.35	0.65	7.50	0.55
14)	*Tegula* sp (TE)	2.0		**35.6**	**4.3**	**14.5**	0.79	0.30	0.37	0.63	7.30	0.59
15)	*Tetrapygus niger* (TN)	2.0		**51.6**	**3.1**	**10.5**	0.62	0.30	0.37	0.63	5.30	0.58
16)	*Leukoma thaca (LT)*	2.1	**0.2**	**19.3**	**3.3**	**11.0**	0.94	0.30	0.38	0.63	5.50	0.60
17)	*Austromegabalanus psittacus* (AP)	2.1	**0.01**	**20.8**	**1.1**	**7.0**	0.04	0.16	0.20	0.80	4.50	0.24
18)	*Pyura chilensis* (PYC)	2.4	**0.12**	**19.1**	**3.2**	**11.0**	0.93	0.29	0.36	0.64	5.60	0.57
19)	*Lessonia trabeculata* (LT)	1.0	**0.06**	**3589.0**	**3.4**	**0.0**	0.96					
20)	*Rhodymenia* sp. (RHO)	1.0		**150.6**	**5.0**	**0.0**	0.88					
21)	Chlorophyta (CH)	1.0		**355.0**	**5.0**	**0.0**	0.97					

(continued)

Table 2 (continued)

	Group/specie name	TL	Ca	B	P/B	Q/B	EE	GE	NE	RA/AS	RA/B	P/RA
22)	Rhodophyta (RH)	1.0		**816.2**	**5.5**	**0.0**	0.99					
23)	Phaeophyta (PHA)	1.0		**1911.0**	**8.5**	**0.0**	0.99					
24)	Zooplankton (ZOO)	2.0		**18.0**	**80.0**	**270.0**	0.77	0.30	0.37	0.63	136.00	0.59
25)	Phytoplankton (PHY)	1.0		**38.0**	**280.0**	**0.0**	0.87					
26)	Detritus (DET)	1		**100.0**			0.72					

from the literature. To determine the food spectra of the principal benthic species, the stomach and gut were extracted and the gut contents were classified to the lowest possible taxonomic level; the frequency of occurrence of each food item was then calculated. During the model balancing procedure, the unassimilated food (%) for the fish *C. variegatus* and *Pinguipes chilensis* was slightly increased, since both species principally convert the energy in somatic tissue from filter feeders (sensu Docmac, Araya, Hinojosa, Dorador, & Harrod, 2017). In this case, filter feeders are integrated as a part of the Small Epifauna Omnivore functional group. The balanced models were checked based on the six guidelines proposed by Heymans et al. (2016): (1) the Ecotrophic Efficiency (EE) of all compartments was < 1.0 (Ricker, 1968) and that (2) the Gross Efficiency (GE) of all compartments was < 0.3 (Christensen & Pauly, 1993). If any inconsistencies were detected, the average biomass was modified within the confidence limits (±1 standard deviation). It was checked that (3) the Net Efficiency of all compartments was larger than the Gross Efficiency, (4) the Respiration/Assimilation Biomass (RA/AS) was < 1.0, (5) the Respiration/Biomass (RA/B) values for fishes were between 1 and 10 $year^{-1}$ and for groups with higher turnover between 50 and 100 $year^{-1}$ and (6) the Production/Respiration (P/RA) was <1.0 (Table 2).

2.3 *Macroscopic Ecosystem Properties*

The following macrodescriptors are based on Ulanowicz's *Ascendency* that enables quantification of the level of development and organisation of ecosystems (Ulanowicz 1986, 1997). (1) *Total Biomass/Total System Throughput* (TB/TST) ratio suggests different states of system maturity (Christensen, 1995); (2) *Total System Throughput* (TST) indicates the size of the system, that is, the total number of flows in the system; (3) *Average Mutual Information* (AMI) quantifies the organisation of the system in relation to the number and diversity of interactions between components (complexity); (4) *Ascendency* (A) measures the growth and development of a system and integrates TST and AMI of flows; (5) *Overhead* (Ov) quantifies the degrees of freedom preserved by the network and can be used to estimate the ability of a network to withstand perturbations (can be estimated as C-A); (6) *Development Capacity* (C), the upper limit of *Ascendency*. As a derived measure, (7) the ratios of A/C and Ov/C are used as indicators of ecosystem development and the ability of the system to resist disturbances (Baird & Ulanowicz, 1993; Costanza & Mageau, 1999; Kaufman & Borrett, 2010). Finally, (8) Relative Internal Ascendency (A_i/C_i) represents well-organised, mature and efficient systems that are resistant to perturbations (Baird, McGlade, & Ulanowicz, 1991; Baird & Ulanowicz, 1993). The algorithms of macroscopic properties are shown in Box 3.

Box 3 Algorithms of Macroscopic Network Properties

Total system throughput (TST) (g m^{-2} year^{-1})	$T = \sum_{i}^{n+1} \sum_{j}^{n+2} T_{ij}$
Average mutual information (AMI)	$\mathrm{AMI} = \sum_{i}^{n+2} \sum_{j}^{n+2} f_{ij} * Q_i * \log_2 \left(\frac{f_{ij}}{Q_j} \right)$
Ascendency (Flowbits)	$A = -T * \sum_{i=0}^{n} Q_i * \log Q - \left[T * \sum_{i=0}^{n+2} \sum_{J=0}^{n+2} f_{ij} * Q_j \log \left(\frac{f_{ij} * Q_i}{Q_i} \right) \right]$
Total capacity (Flowbits)	$C = A + O$
Overhead (Flowbits)	$O = C - A.$

2.4 *Determination of Keystone Species*

Functional Index

We used the functional keystoneness index (KS_i) proposed by Libralato et al. (2006), which is an extension of *Mixed Trophic Impact* (MTI) (Ulanowicz & Puccia, 1990) to measure the total effect of each component:

$$\varepsilon_i = \sqrt{\sum_{j \neq i}^{n} m_{ij}^2}, \tag{8}$$

where m_{ij} represents the elements of the MTI matrix and quantifies the direct and indirect impacts that each group i has on any group j of the food web. However, the effect of the change in biomass on the group itself (m_{ij}) is not included. The contribution of the biomass of each group was estimated, with respect to the total biomass of the food web using the following equation:

$$p_i = \frac{B_i}{\sum_{i}^{n} B_k}, \tag{9}$$

where p_i is the biomass proportion of each species B_i with respect to the sum of the total biomass B_k. Therefore, to balance the overall effect and the biomass, we established the keystone index KS_i for each species or functional group by integrating Eqs. (8) and (9) as follows:

$$\mathrm{KS_i} = \log\left[\varepsilon_i\left(1 - p_i\right)\right]. \tag{10}$$

This index assigns high keystone species values to those species or functional groups with low biomass and high overall effects.

In addition, the outcomes of the propagation of direct and indirect effects and the magnitudes of the *System Recovery Time* (SRT) that were estimated by *Ecosim* were analysed with Eqs. (8)–(10) to obtain two additional functional keystone indices. The *Ecosim* simulations were used to evaluate the propagation of instantaneous direct and indirect effects and SRT (as a system resilience measure) in response to an increase in the total mortality (Z) of all compartments [see Eqs. (11) and (12)], which was set equivalent to 10, 30 and 50%.

$$Z = M\left(\text{natural mortality}\right) + F\left(\text{fishing mortality}\right) \tag{11}$$

$$\text{Production}\left(P\right) = \text{Biomass}\left(B\right) * Z. \tag{12}$$

This procedure was performed between the first and second years of the simulation for all components considered in the model. These three Z magnitudes were set for prediction purposes as a measure of confidence (sensu Ortiz et al., 2015). Since the model represented only short-term dynamics, the propagation of instantaneous effects was determined by evaluating the changes in biomass of the remaining variables during the third simulation year. All dynamic simulations by *Ecosim* were carried out using the following vulnerabilities (flow control) (ν_{ij}): (1) bottom-up (prey control the flow), (2) top-down (predators control the flow) and (3) mixed (both preys and predators control the flow).

Equations (8)–(10) were used to obtain one keystone species index that was related to the propagation of direct and indirect effects ($\mathrm{KS_{iEcosim1}}$); Equations (9) and (10) were used to obtain another functional keystone index related to the SRT values ($\mathrm{KS_{iEcosim2}}$). Both indices and the $\mathrm{KS_i}$ index (Libralato et al., 2006) revealed that high keystoneness values corresponded to the species and functional groups with low biomasses and high overall effects.

Topological or Structural Index

The topological index (K_i) proposed by Jordán, Takács-Sánta, and Molnár (1999) and Jordán, (2001) considers direct and indirect interactions in bottom-up (K_b) and top-down (K_t) directions. This index is calculated as follows:

$$K_i = \sum_{c=1}^{n} \frac{1}{\mathrm{d_c}}\left(1 + K_{\mathrm{bc}}\right) + \sum_{e=1}^{n} \frac{1}{\mathrm{f_e}}\left(1 + K_{\mathrm{te}}\right), \tag{13}$$

where n is the number of predators eating species i, d_c is the number of prey of the cth predator, K_{bc} is the bottom-up keystone index of the cth predator, and symmetrically: m is the number of prey eaten by species i, f_e is the number of predators of the eth prey and K_{te} is the top-down keystone index for the eth prey. Within this index, the first and second components represent the bottom-up (K_{bc}) and top-down (K_{te}) effects, respectively. Finally, the keystone index K_i assigns high keystone species values to the compartments that are greatly affected by a perturbation of the network. For more details on this method, see Jordán (2001) and Vasas, Lancelot, Rousseau, and Jordán, (2007). Only the bottom-up and top-down components of K_i were used in the current work as a way to compare the functional indices obtained from the *Ecosim* simulations under comparable flow control mechanisms.

Semi-Quantitative Keystone Index

Keystoneness indices based on qualitative loop models were also calculated. Once the matrix stabilised at $F_n < 0$, the self-dynamics of each variable were modified to estimate a new perturbed magnitude of the local stability (F_p) based on the distance (difference, Δ) between F_n and F_p, as shown in Eq. (14):

$$\Delta = \left| F_n - F_p \right|. \tag{14}$$

This index allowed a change to the initial local stability (F_n) that was provoked by each variable, thereby obtaining a semi-qualitative keystone species index ($KQ_{i\ LA1}$). Since *Loop Analysis* does not quantify the abundance of the variables, the difference (Δ) was treated similarly as in Eq. (10) to obtain an additional keystone index ($KQ_{i\ LA2}$), in which high keystoneness values corresponded to variables with low biomass and high overall effects. Due to the qualitative characteristics of *Loop Analysis*, the predator–prey interaction was captured as a mixed control mechanism.

Centrality of Nodes Sets

The software Key-Player 1.45 (Borgatti, 2003) was used to compute the importance of species combinations in maintaining the integrity of a network. The importance of a set of nodes can be calculated by considering either their fragmentation effect (KPP1) or their reachability effect (KPP2). In the first case (F), it is identified which k nodes should be deleted from the network of n nodes in order to maximally increase its fragmentation. In the second case (R), it is identified from which k nodes the largest proportion of the other n-k nodes are reachable within a certain distance. Based on fragmentation (F of KPP1), the best set of the deleted k nodes can maximally increase the fragmentation of the network. This means an increase to the number of components and a larger average distance generated within individual components. We used k = 1, 2 and 3 with 10,000 simulations for each. We also considered the distance-based reachability approach (*Rd* of KPP2). We counted the number of nodes that were reachable within a given distance of $m = 1$ step from a given set of k nodes. We chose $m = 2$ steps and increased the size of the KP-set from $k = 1$ to $k = 3$ and we applied 10,000 runs for each simulation. The outcome was three sets of nodes (for $k = 1, 2, 3$) for each network, containing species codes. For each k, the software presents the percentage of nodes outside the KP-set but reachable from it in l step. If this percentage reaches 100%, then the whole network is reachable from the KP-set and we cannot create larger KP-sets.

3 Macroscopic Ecosystem Properties and Keystone Species Complex

We characterise the studied systems first by the biomass of some individual species, then by systemic indicators. This study shows that sites characterised by low and middle human disturbance had the greatest biomass of *Lessonia trabeculata* (kelp); the sites with higher human disturbance (shipping, fishing and tourism) showed higher biomass of the gastropod *Turritella cingulata* and the algae *Gracilaria* sp. and *Ulva* sp. (Table 2). These outcomes demonstrate the changes the structure, organisation and performance of the ecosystem under different disturbance regimes (González et al., 2016; Pauly et al., 1998; Petersen et al., 2008; Ray et al., 2000). The kelp forests are considered an important economic source for coastal human

Table 3 Summary statistics after mass-balanced process by *Ecopath* and network flow indices for benthic ecological systems of bays: Guala-Guala (GGB), Escondida (EB), El Cobre (ECB), El Blanco (EBB), Caldera (CB), Inglesa (IB) and Salado (SB). The units are in g wet weight, bits is a unit of information and Flowbits is the product of flow (g ww m^{-2} $year^{-1}$)

Parameter	GGB	EB	ECB	EBB	CB	IB	SB
Total system throughput (TST) (g ww m^{-2} $year^{-1}$)	94,961	100,092	48,072	64980	863,136	148,054	150,336
Total biomass/Total system throughput (TB/TST)	0.083	0.094	0.073	0.088	0.074	0.085	0.077
Ascendency (A) (Total) Flowbits	138,862	146,331	55,266	86,746	1,035,067	222,049	178,707
Overhead (Ov) (Total) Flowbits	308,657.50	279,489	186,698	212,526	2,646,337	370,814	401,525
Development capacity (C) (total) Flowbits	450,120.70	428,685	247,425	299,916	3,681,411	592,869	580,233
Ov/C (%)	0.69	0.65	0.75	0.71	0.72	0.63	0.69
A/C (%)	0.31	0.34	0.22	0.29	0.28	0.37	0.31
A_i/C_i (internal) (%)	0.25	0.27	0.17	0.21	0.24	0.28	0.27
Average mutual information (AMI) (bits)	1.46	1.46	1.14	1.33	1.19	1.49	1.18

communities but have been heavily exploited, causing drastic changes in the intertidal and subtidal benthic communities at temporal and spatial scales (Ortiz & Levins, 2017; Vásquez et al., 2013). The gastropod *T. cingulata* is considered a common biological component in disturbed areas due to its high growth rate and resistance against eutrophic environments. Likewise, *T. cingulata* dominance may alter the structure of the communities (Cardoso, Pardal, Raffaelli, Baeta, & Marques, 2004; Cummins, Roberts, & Zimmerman, 2004; Martins, Pardal, Lillebo, Flindt, & Marques, 2001). Considering the responses of individual species, we may not see all relevant changes in the ecosystem, so we also apply systemic indicators.

Inglesa and Salado bays should be considered the most developed, organised and healthy ecosystems based on *Ascendency*, TST, Ov, C, A/C, Ai/Ci and AMI (Table 3). These bays do not present intense human disturbances, especially compared to Caldera bay; therefore, ecosystem functioning has not yet been intensely disturbed. High values of TST may be a consequence of the high biomass of *L. trabeculata* in the Salado bay, which encourages a high rate of cycling and greater amount of flows. Inglesa bay was previously considered a highly disturbed site; however, the impact caused by the seasonal tourism on ecosystem properties seems to be low, which is reflected in the high values of the macroscopic descriptors. Likewise, the magnitudes of A, C, A/C and AMI were higher than other ecological systems in northern Chile, such as Mejillones Peninsula (Ortiz, 2010; Ortiz et al., 2015), Marine Reserve La Rinconada (Ortiz, Avendaño, Cantillañez, Berrios, & Campos, 2010) and Tongoy Bay (González et al., 2016; Ortiz & Wolff, 2002; Wolff, 1994). Even though Caldera bay exhibited the highest magnitudes for growth and development compared to the other ecosystems (explained by its high values of TST), its low values of organisation (measured as AMI) demonstrate features of enrichment or eutrophication (sensu Ulanowicz, 1997). Caldera bay supports severe anthropogenic disturbances, such as mining industries, ports, fishing and farming activities.

The single-species keystone indices indicated that the keystone species complexes (KSCs) are composed of several species and functional groups located at different trophic levels and linked trophically (Fig. 1). Our findings also showed that different species of carnivores formed the KSCs, highlighted by the starfish *M. gelatinosus*, which integrated the majority of the KSCs. This result agrees with Paine, Castilla, and Cancino (1985) and Gaymer and Himmelman (2008), who referred to *M. gelatinosus* as a keystone species in subtidal benthic communities along the Chilean coast. Likewise, Ortiz, Campos, et al. (2013), Ortiz, Levins, et al. (2013) indicated that this species is part of the KSCs in subtidal and intertidal ecological systems in northern Chile. Similarly, the small epifauna omnivore functional group was present in the most of the KSCs; therefore, these species also play an important role by providing food for a variety of components (Bradshaw, Collins, & Brand, 2003; Gili & Hughes, 1995; Taylor, 1998).

Other carnivore species were also part of the KSCs, such as the coastal fish *C. variegatus* and the crab *Romaleon setosum* (a prominent benthic predator), which agree with the outcomes previously described by Ortiz, Campos, et al. (2013) and González et al. (2016). Additionally, González et al. (2016) showed that the exploitation of *R. setosum* would propagate high quantitative impacts on the other ecosystem compartments. Finally, the cephalopod *Octopus* sp. also plays a role in the KSCs because they occupy a central-top position in marine trophic networks; they are prey for marine mammals and seabirds and predators to crustaceans and fishes (Cortez, Castro, & Guerra, 1995; Klages, 1996; Leite, Haimovici, & Mather, 2009; Xavier & Croxall, 2007).

The main herbivores that integrated the KSCs was the sea urchin *T. niger*. This outcome also agrees with Ortiz, Campos, et al. (2013), who reported that this sea urchin is a part of the KSCs in kelp forests of benthic ecosystems of Mejillones Peninsula (northern Chile). This sea urchin is a conspicuous benthic grazer; it is one of the most abundant grazers along the north-central coast of Chile and dominates in the barrens (Steneck et al., 2002; Tegner & Dayton, 2000; Uribe, Ortiz, Macaya, & Pacheco, 2015; Vásquez & Buschmann, 1997). Therefore, changes in abundance of this sea urchin and the intensive exploitation of kelp could eventually have synergetic negative impacts on the kelp forest ecosystem along the north-central Chilean coast (Ortiz, 2003; Ortiz & Levins, 2017; Vásquez, 2008; Vásquez et al., 2013). Likewise, herbivore species such as the gastropods *T. cingulata* and *Tegula* sp. were also included in the KSCs; both species have previously been described to have keystoneness properties along the northern Chilean coast (González et al., 2016; Ortiz, Campos, et al., 2013).

The keystone role of the phytoplankton functional group is due to the coastal upwelling that influences the primary productivity of subsurface waters; this high concentration of phytoplankton and the almost automatically generated central position in the interaction network (Escribano, Rosales, & Blanco, 2004; Marín, Rodríguez, Vallejo, Fuenteseca, & Oyarce, 1993). Likewise, the centrality of the detritus functional group is an artefact: it is due to the high level of aggregation of this group and the rich trophic relationships it has. Apart from being a sink to all dead material, several studies have emphasised the importance of bacteria as food for various species of molluscs (Grossmann & Reichardt, 1991; Plante & Mayer, 1994; Plante & Shriver, 1998), zooplankton (Epstein, 1997) and Echinodermata (Findlay & White, 1983). Most of the KSCs concentrated the lowest magnitudes of the total system biomass (Fig. 1). Only the KSC of Caldera bay represented up to 48% of the total system biomass using functional, topological and semi-quantitative indices, which can be explained by the high abundance of the snail *T. cingulata*. A putative explanation is that Caldera bay is highly impacted by different human activities, especially the subtidal farming of macroalgae, which increases the food availability for this snail.

Based on the extended concept of KSC, the species and functional groups that defined the complexes are characterised by their low abundances and high impacts on the ecological system. In this sense, one would expect the KSC to show low values of TST and AMI. However, the outcomes do not agree entirely with this tendency. The average TST and AMI were 27.7 % and 33.3%, respectively (Table 4). These high values could be explained by the presence of species and functional groups such as zooplankton, phytoplankton, detritus and *T. cingulata* that provide high flows (TST) and high values of AMI. If these components are excluded from

Table 4 Contribution of each component of KSCs on the emergent properties in the study sites along the northern Chilean coast. *B* biomass, *TST* total system throughput, *AMI* average mutual information, *A* ascendency, *Ov/C* overhead/capacity ratio. The values in parenthesis indicate the magnitude of emergent properties when Phytoplankton, Phaeophyta and *Turritella cingulate* are excluded

(a) GGB	B (%)	TST (%)	AMI (%)	A (%)	Ov/C (%)
Meyenaster gelatinosus	0.2	0.07	0.07	0.07	4.63
Octopus sp.	0.4	0.28	0.16	0.16	4.81
Small epifauna omnivore	5.3	6.41	6.24	6.24	4.01
Tetrapygus niger	1.0	0.92	1.25	1.25	4.09
Phaeophyta	5.5	3.90	7.11	7.10	3.31
Phytoplankton	0.6	14.74	19.02	19.03	3.09
Total	**13.0**	**26.32**	**33.86**	**33.85**	**23.93**
		(11.58)	**(14.84)**	**(14.82)**	
(b) EB					
Meyenaster gelatinosus	0.1	0.04	0.05	0.05	4.65
Heliaster helianthus	0.4	0.17	0.14	0.14	4.81
Romaleon setosum	0.3	0.29	0.23	0.23	4.78
Small epifauna omnivore	3.1	4.28	4.76	4.76	4.10
Tegula sp.	3.7	5.07	5.06	5.06	3.99
Phaeophyta	2.9	1.35	2.82	2.82	3.39
Phytoplankton	0.5	13.99	19.63	19.61	3.03
Total	**11.0**	**25.18**	**32.69**	**32.68**	**28.74**
		(11.20)	**(13.06)**	**(13.07)**	
(c) ECB					
Meyenaster gelatinosus	0.2	0.07	0.07	0.07	5.15
Romaleon setosum	0.4	0.31	0.28	0.28	5.13
Cheilodactylus variegatus	0.2	0.14	0.13	0.13	5.18
Small epifauna omnivore	9.5	10.27	9.91	9.88	4.51
Gracilaria sp.	5.5	3.53	1.79	1.79	3.01
Phytoplankton	0.8	16.31	21.22	21.22	3.80
Total	**16.7**	**30.61**	**33.40**	**33.37**	**26.80**
		(14.31)	**(12.18)**	**(12.15)**	
(d) EBB					
Meyenaster gelatinosus	0.1	0.03	0.03	0.03	4.43
Romaleon setosum	0.3	0.27	0.30	0.30	4.17
Small epifauna omnivore	3.4	4.45	5.25	5.25	3.75
Tetrapygus niger	0.6	0.52	0.84	0.84	3.82
Loxechinus albus	0.2	0.19	0.16	0.16	4.44
Phaeophyta	4.4	3.32	5.43	5.43	3.31
Phytoplankton	0.5	12.07	15.66	15.66	3.19
Total	**9.4**	**20.85**	**27.66**	**27.67**	**27.11**
		(8.79)	**(12.01)**	**(12.01)**	
(e) CB					

(continued)

Table 4 (continued)

(a) GGB	B (%)	TST (%)	AMI (%)	A (%)	Ov/C (%)
Octopus sp.	0.001	0.00058	0.0005	0.0004	4.58
Pinguipes chilensis	0.002	0.0010	0.0015	0.0015	4.35
Romaleon setosum	0.01	0.0060	0.0057	0.0056	4.48
Small epifauna omnivore	3.8	4.73	5.47	5.47	3.76
Turritella cingulata	29.9	32.26	29.86	29.81	2.83
Phaeophyta	12.4	7.87	5.66	5.66	4.01
Phytoplankton	0.3	6.65	10.34	10.33	3.31
Total	**46.480**	**51.52**	**51.34**	**51.28**	**27.32**
		(19.26)	**(21.48)**	**(21.47)**	
(f) IB					
Meyenaster gelatinosus	1.5	0.63	0.62	0.62	4.29
Pinguipes chilensis	0.0	0.02	0.01	0.01	4.73
Small epifauna carnivore	1.2	1.29	1.43	1.43	4.18
Tegula sp.	0.9	1.09	1.38	1.38	4.06
Austromegabalanus psittacus	0.4	0.23	0.30	0.30	4.22
Phaeophyta	4.3	3.16	6.32	6.32	2.74
Zooplankton	0.1	3.45	4.05	4.05	3.79
Phytoplankton	0.2	5.82	12.00	12.01	2.63
Total	**8.7**	**15.69**	**26.11**	**26.13**	**30.64**
		(9.87)	**(14.11)**	**(14.12)**	
(g) SB					
Meyenaster gelatinosus	0.1	0.02	0.02	0.02	4.20
Small epifauna omnivore	2.2	2.43	3.17	3.18	3.65
Tegula sp.	0.3	0.34	0.40	0.40	4.00
Austromegabalanus psittacus	0.2	0.10	0.09	0.09	4.16
Phaeophyta	16.4	10.81	11.44	11.42	3.00
Zooplankton	0.2	3.23	3.36	3.36	3.79
Phytoplankton	0.3	7.08	9.84	9.87	3.33
Total	**19.6**	**24.01**	**28.33**	**28.35**	**26.12**
		(13.21)	**(16.90)**	**(16.93)**	
Total average		**27.74**	**33.34**	**33.33**	**27.24**
		(14.11)	**(17.50)**	**(17.50)**	

the KSCs, the values are significantly lower and achieve an approximate total average of 18 % (Table 4).

The small epifauna omnivore functional group (SEO) is the most frequent common element of the KSCs described according to the two different methods (in 5 out of 7 models). SEO is missing in the two mostly disturbed systems (Caldera bay, Inglesa bay). It can be accompanied also by the starfish *M. gelatinosus* (MG) or phytoplankton (PHY). If SEO is not present among the common elements, it is either PHY or Octopus (OC) is the intersection between the KSCs. Octopus is a common KSC element only in Caldera bay, where KSC represents a high biomass in the ecosystem. In general, the top predator included in the KSCs (such as *M. gelatinosus*, *Octopus* sp. and *Pinguipes chilensis*) exhibited low values for AMI and high values for Ov/C, which indicates that these species contribute simultaneously to the complexity and the resistance capacity of the systems.

4 Conclusions and Future Directions

The macroscopic indices determined for the seven trophic ecosystem models improved our understanding about of the structure and dynamics of the benthic system. Different ecosystem properties indicate that sites far from human settlements and/or under low disturbances would be more developed, organised, mature and healthy ecological systems. Therefore, these sites could be candidates for monitoring programmes in order to evaluate the trajectory of ecosystem health and conservation. In addition, it is essential to evaluate the trajectory of exploited species within an ecosystemic context; these species constitute compartments with relevant roles in the trophic structure and functioning of kelp forest ecosystems. Likewise, we believe that the keystone species complex (KSC) can facilitate the design and assessment of conservation and monitoring measures by selecting producers, intermediate consumers and top carnivores at the same time. This is particularly relevant in kelp forest ecosystems, which are being severely stressed by the direct effects of fisheries, pollution and tourism. Keystone species complexes have emerged as a generality in ecological systems along the Chilean coast.

Acknowledgments The information used for this study was obtained via the grants INNOVA-CORFO 05CR11IXM-03 (Región de Atacama, Chile) and 09CN14-5873 (Región de Antofagasta, Chile). Part of this work corresponds to the doctoral thesis of first author, funded by a scholarship of the Instituto Antofagasta, Universidad de Antofagasta, Chile.

References

Allen, K. R. (1971). Relation between production and biomass. *Journal of the Fisheries Research Board of Canada, 28*, 1573–1581.

Almunia, J., Basterretxea, G., Aristegui, J., & Ulanowicz, R. E. (1999). Benthic-pelagic switching in a Coastal subtropical lagoon. *Estuarine, Coastal and Shelf Science, 49*, 363–384.

Baird, D., McGlade, J. J., & Ulanowicz, R. E. (1991). The comparative ecology of six marine ecosystems. *Philosophical Transactions of the Royal Society B: Biological Sciences, 333*, 15–29.

Baird, D., & Ulanowicz, R. (1993). Comparative study on the trophic structure, cycling and ecosystem properties of four tidal estuaries. *Marine Ecology Progress Series, 99*, 221–237.

Benerjee, A., Scharler, U. M., Fath, B. D., & Ray, S. (2017). Temporal variation of keystone species and their impact on system performance in a South African estuarine ecosystem. *Ecological Modelling, 363*, 207–220.

Borgatti, S. P. (2003). *Key Player*. Boston: Analytic Technologies.

Bradshaw, C., Collins, P., & Brand, A. R. (2003). To what extend does upright sessile epifauna affect benthic biodiversity and community composition? *Marine Biology, 143*, 783–791.

Cardoso, P. G., Pardal, M. A., Raffaelli, D., Baeta, A., & Marques, J. C. (2004). Macroinvertebrates response to different species of macroalgal mats and the role of disturbance history. *Journal of Experimental Marine Biology and Ecology, 308*, 207–220.

Caro, T. (2010). *Conservation by proxy: indicator, umbrella, keystone, flagship and other surrogate species*. New York: Island Press.

Cerda, G., & Wolff, M. (1993). Feeding ecology of the crab Cancer polyodon in La Herradura Bay, northern Chile. II. Food spectrum and prey consumption. *Marine Ecology Progress Series, 100*, 119–125.

Christensen, V. (1995). Ecosystem maturity–towards quantification. *Ecological Modelling, 77*, 3–32.

Christensen, V., & Pauly, D. (1993). Trophic models in aquatic ecosystem. *ICLARM Conference Proceedings, 26*, 390.

Christensen, V., & Walters, C. J. (2004). Ecopath with Ecosim: methods, capabilities and limitations. *Ecological Modelling, 172*, 109–139.

Cortez, T., Castro, B. G., & Guerra, A. (1995). Feeding dynamics of *Octopus mimus* (Mollusca: Cephalopoda) in northern Chile waters. *Marine Biology, 123*, 497–503.

Costanza, R. (1992). Toward an operational definition of health. In R. Costanza, B. Norton, & B. Haskell (Eds.), *Ecosystem health: New goals for environmental management*. Washington: Island Press.

Costanza, R., & Mageau, M. (1999). What is a healthy ecosystem? *Aquatic Ecology, 33*, 105–115.

Cummins, S. P., Roberts, D. E., & Zimmerman, K. D. (2004). Effects of the green macroalga *Enteromorpha intestinalis* on macrobenthic and Seagrass assemblages in a shallow coastal estuary. *Marine Ecology Progress Series, 266*, 77–87.

Daily, G. C., Ehrlich, P. R., & Haddad, N. M. (1993). Double keystone bird in a keystone species complex. *Proceedings of the National Academy of Sciences USA, 90*, 592–594.

Docmac, F., Araya, M., Hinojosa, I. A., Dorador, C., & Harrod, C. (2017). Habitat coupling writ large: Pelagic-derived materials fuel benthivorous macroalgal reef fishes in an upwelling zone. *Ecology, 98*, 2267–2272.

Epstein, S. (1997). Microbial food web in marine sediments: I. Trophic interactions and grazing rates in two tidal flat communities. *Microbial Ecology, 34*, 188–198.

Escribano, R., Rosales, S. A., & Blanco, J. L. (2004). Understanding upwelling circulation off Antofagasta (northern Chile): A three-dimensional numerical-modeling approach. *Continental Shelf Research, 24*, 37–53.

Estes, J. A., & Palmisano, J. F. (1974). Sea otters: their role in structuring nearshore communities. *Science, 185*, 1058–1060.

Estes, J. A., Tinker, M. T., Williams, T. M., & Doak, D. F. (1998). Killer whale predation on sea otters liking oceanic and nearshore ecosystems. *Science, 282*, 473–476.

Findlay, R., & White, D. (1983). The effects of feeding by the sand dollar *Mellita quinquiesperforata* (Leske) on the benthic microbial community. *Journal of Experimental Marine Biology and Ecology, 72*, 25–41.

Gaymer, C. F., & Himmelman, J. H. (2008). A keystone predatory sea star in the intertidal zone is controlled by a higher-order predatory sea star in the subtidal zone. *Marine Ecology Progress Series, 370*, 143–153.

Giacaman-Smith, J., Neira, S., & Arancibia, H. (2016). Community structure and trophic interactions in a coastal management and exploitation area for benthic resources in central Chile. *Ocean and Coastal Management, 119*, 155–163.

Gili, J. M., & Hughes, R. G. (1995). The ecology of marine benthic hydroids. *Oceanography and Marine Biology, 33*, 351–426.

González, J., Ortiz, M., Rodríguez-Zaragoza, F., & Ulanowicz, R. (2016). Assessment of long-term changes of ecosystem indexes in Tongoy Bay (SE Pacific coast): Based on trophic network analysis. *Ecological Indicators, 69*, 390–399.

Grossmann, S., & Reichardt, W. (1991). Impact of *Arenicola marina* on bacteria in intertidal sediments. *Marine Ecology Progress Series, 77*, 85–93.
Halpern, B. S., Walbridge, S., Selkoe, K. A., Kappel, C. V., Micheli, F., D'Agrosa, C., Bruno, J. F., Casey, K. S., Ebert, C., Fox, H. E., Fujita, R., Heinemann, D., Lenihan, H. S., Madin, E. M. P., Perry, M. T., Selig, E. R., Spalding, M., Steneck, R., & Watson, R. (2008). A global map of human impact on marine ecosystems. *Science, 319*, 948–952.
Hermosillo-Núñez, B. B., Ortiz, M., Rodríguez-Zaragoza, F. A., & Cupul-Magaña, A. L. (2018). Trophic network properties of coral ecosystems in three marine protected areas along the Mexican Pacific Coast: Assessment of systemic structure and health. *Ecological Complexity, 36*, 73–85.
Heske, E. J., Brown, J. H., & Mistry, S. (1994). Long-term experimental study of a Chihuahuan desert rodent community: 13 years of competition. *Ecology, 75*, 438–445.
Heymans, J. J., Coll, M., Link, J. S., Mackinson, S., Steenbeek, J., Walters, C., & Christensen, V. (2016). Best practice in Ecopath with Ecosim food-web models for ecosystem-based management. *Ecological Modelling, 331*, 173–184.
Jackson, J. B. C., Kirby, M., Berger, W. H., Bjorndal, K. A., Botsford, L. W., Bourque, B. J., Bradbury, R. H., Cooke, R., Erlandson, J., Estes, J. A., Hughes, T. P., Kidwell, S., Lange, C., Lenihan, H. S., Pandolfi, J. M., Peterson, C. H., Steneck, R. S., Tegner, M. J., & Warner, R. R. (2001). Historical overfishing and the recent collapse of coastal ecosystems. *Science, 293*, 629–637.
Jones, C. G., Lawton, J. H., & Shachak, M. (1994). Organisms as ecosystem engineers. *Oikos, 69*, 373–386.
Jordán, F. (2001). Trophic fields. *Community Ecology, 2*, 181–185.
Jordán, F., & Molnár, I. (1999). Reliable flows and preferred patterns in food webs. *Evolutionary Ecology Research, 1*, 591–609.
Jordán, F., Okey, T. A., Bauer, B., & Libralato, S. (2008). Identifying important species: Linking structure and function in ecological networks. *Ecological Modelling, 216*, 75–80.
Jordán, F., Pereira, J., & Ortiz, M. (2019). Mesoscale network properties in ecological system models. *Current Opinion in Systems Biology, 13*, 122–128.
Jordán, F., Takács-Sánta, A., & Molnár, I. (1999). A reliability theoretical quest for keystones. *Oikos, 86*, 453–462.
Kaufman, A. G., & Borrett, S. R. (2010). Ecosystem network analysis indicators are generally robust to parameter uncertainty in a phosphorus model of Lake Sideney Lanier, USA. *Ecological Modelling, 221*, 1230–1238.
Klages, N. (1996). Cephalopods as prey. II. Seals. *Philosophical Transactions of the Royal Society B: Biological Sciences, 351*, 1045–1052.
Laland, K. N., Odling-Smee, F. J., & Feldman, M. W. (1996). The evolutionary consequences of niche construction: a theoretical investigation using two-locus theory. *Journal of Evolutionary Biology, 9*, 293–316.
Leite, T. S., Haimovici, M., & Mather, J. (2009). *Octopus insularis* (Octopodidae), evidences of a specialized predator and a time-minimizing hunter. *Marine Biology, 156*, 2355–2367.
Levins, R. (1968). Evolution in changing environments. Princeton Monograph Series.
Levins, R. (1974). The qualitative analysis of partially specified systems. *Annals of the New York Academy of Sciences, 231*, 123–138.
Levins, R. (1998). Qualitative mathematics for understanding, prediction, and intervention in complex ecosystems. In D. Raport, R. Costanza, P. Epstein, C. Gaudet, R. Levins (Eds.), *Ecosystem Health, Blackwell Science* (pp. 178–204). MA.
Levins, R., & Lewontin, R. C. (1985). *The dialectical biologist*. Cambridge: Harvard University Press.
Libralato, S., Christensen, V., & Pauly, D. (2006). A method for identifying keystone species in food web models. *Ecological Modelling, 195*, 153–171.
Lotze, H. K., Lenihan, H. S., Bourque, B. J., Bradbury, R. H., Cooke, R. G., Kay, M. C., Kidwell, S. M., Kirby, M. X., Peterson, C. H., & Jackson, J. B. C. (2006). Depletion, degradation, and recovery potential of estuaries and coastal seas. *Science, 312*, 1806–1809.

Marín, V., Rodríguez, L., Vallejo, L., Fuenteseca, J., & Oyarce, E. (1993). Efectos de la surgencia costera sobre la productividad primaria primaveral de Bahía Mejillones del Sur (Antofagasta, Chile). *Revista Chilena de Historia Natural, 66*, 479–491.

Martins, I., Pardal, M. Â., Lillebo, A. I., Flindt, M. R., & Marques, J. C. (2001). Hydrodynamics as a major factor controlling the occurrence of green macroalgal blooms in a eutrophic estuary: A case study on the influence of precipitation and river management. *Estuarine, Coastal and Shelf Science, 52*, 165–177.

Monaco, M., & Ulanowicz, R. (1997). Comparative ecosystem trophic structure of three U.S. mid-Atlantic estuaries. *Marine Ecology Progress Series, 161*, 239–254.

Noss, R. F. (1999). Assessing and monitoring forest biodiversity: A suggested framework and indicators. *Forest Ecology and Management, 115*, 135–146.

Odum, E. P. (1969). The strategy of ecosystem development. *Science, 164*, 262–270.

Odum, H. T., & Odum, E. P. (1955). Trophic structure and productivity of a Windward Coral Reef Community on Eniwetok Atoll. *Ecological Monographs, 25*, 291–320.

Okey, T. (2004). *Shifted community states in four marine ecosystems: some potential mechanisms*. PhD Thesis, The University of British Columbia, Canada, p. 173.

Ortiz, M. (2003). Qualitative modelling of the kelp forest of *Lessonia nigrescens* Bory (Laminariales: Phaeophyta) in eulittoral marine ecosystems of the south–east Pacific: An approach to management plan assessment. *Aquaculture, 220*, 423–436.

Ortiz, M. (2010). Dynamic and spatial models of kelp forest of *Macrocystis integrifolia* and *Lessonia trabeculata* (SE Pacific) for assessment harvest scenarios: Short-term responses. *Aquatic Conservation: Marine and Freshwater Ecosystems, 20*, 494–506.

Ortiz, M., Avendaño, M., Cantillañez, M., Berrios, F., & Campos, L. (2010). Trophic mass balanced models and Dynamic simulations of benthic communities from La Rinconada Marine Reserve off Northern Chile: Network properties and multispecies harvest scenario assessments. *Aquatic Conservation: Marine Freshwater Ecosystem, 20*, 58–73.

Ortiz, M., Berrios, F., Campos, L., Uribe, R., Ramírez, A., Hermosillo-Nuñez, B., González, J., & Rodríguez-Zaragoza, F. (2015). Mass balanced trophic models and short–term dynamical simulations for benthic ecological system of Mejillones and Antofagasta bays (SE Pacific): Comparative network structure and assessment of human impacts. *Ecological Modelling, 309*, 153–162.

Ortiz, M., Campos, L., Berrios, F., Rodríguez, F., Hermosillo, B., & González, J. (2013). Network properties and keystoneness assessment in different intertidal communities dominated by two ecosystem engineer species (SE Pacific coast): A comparative analysis. *Ecological Modelling, 250*, 307–318.

Ortiz, M., Hermosillo-Nuñez, B., González, J., Rodríguez-Zaragoza, F., Gómez, I., & Jordán, F. (2017). Quantifying keystone species complex: Ecosystem-based conservation management in the King George Island (Antarctic Peninsula). *Ecological Indicators, 81*, 453–460.

Ortiz, M., & Levins, R. (2017). Self-feedbacks determine the sustainability of human interventions in eco-social complex systems: Impacts on biodiversity and ecosystem health. *PLoS ONE, 12*(4), e0176163. https://doi.org/10.1371/journal.pone.0176163.

Ortiz, M., Levins, R., Campos, L., Berrios, F., Campos, F., Jordán, F., Hermosillo, B., González, J., & Rodríguez, F. (2013). Identifying keystone trophic groups in benthic ecosystems: Implications for fisheries management. *Ecological Indicators, 25*, 133–140.

Ortiz, M., & Wolff, M. (2002). Trophic models of four benthic communities in Tongoy Bay (Chile): Comparative analysis and preliminary assessment of management strategies. *Journal of Experimental Marine Biology and Ecology, 268*, 205–235.

Paine, R., Castilla, J. C., & Cancino, J. (1985). Perturbation and recovery pattern of starfish-dominated intertidal assemblages in Chile, New Zeland, and Washington State. *American Naturalist, 125*, 679–691.

Paine, R. T. (1969). A note of tropic complexity and community stability. *American Naturalist, 103*, 91–93.

Paine, R. T. (1974). Intertidal community structure. Experimental studies on the relationship between a dominant competitor and its principal predator. *Oecologia, 15*, 93–120.

Pauly, D., Christensen, V., Dalsgaars, J., Froese, R., & Torres, F., Jr. (1998). Fishing down marine food webs. *Science, 279*, 860–863.

Petersen, J. K., Hansen, J. W., Laursen, M. B., Clausen, P., Carstensen, J., & Conley, D. J. (2008). Regime shift a coastal marine ecosystem. *Ecological Applications, 18*, 497–510.

Plante, C., & Mayer, L. (1994). Distribution and efficiency of bacteriolysis in the gut of Arenicola marina and three additional deposit feeders. *Marine Ecology Progress Series, 109*, 183–194.

Plante, C., & Shriver, A. (1998). Patterns of differential digestion of bacteria in deposit feeders: a test of resource partitioning. *Marine Ecology Progress Series, 163*, 253–258.

Ray, S., Ulanowicz, R. E., Majee, N., & Roy, A. (2000). Network analysis of a benthic food web model of a partly reclaimed island in the Sundarban mangrove ecosystem, India. *Journal of Biological Systems, 8*, 263–278.

Ricker, W. E. (1968). Food from the sea. In *Committee on resources and man (Ed). Resource and Man* (pp. 87–108). San Francisco, CA: US National Academy of Sciences, E. H. Freeman.

Simberloff, D. (1998). Flagships, umbrellas, and keystones: Is single-species management passe in the landscape era? *Biological Conservation, 83*, 247–257.

Sparre, P., & Venema S.C. (1997). Introduction to tropical fish stock assessment. Part 1. Manual. In: FAO Fisheries Technical Paper, No. 306. FAO, UN.

Steneck, R. S., Graham, M. H., Bourque, B. J., Corbett, D., Erlandson, J. M., Estes, J. A., & Tegner, M. J. (2002). Kelp forest ecosystems: biodiversity, stability, resilience and future. *Environmental Conservation, 29*, 436–459.

Strub, P., Mesías, J., Montecinos, V., Rutllant, J., & Marchant, S. (1998). Coastal oceanic circulation off western South America. In A. Robinson & K. Birnk (Eds.), *The sea* (pp. 273–314). New York: Wiley.

Taylor, R. B. (1998). Density, biomass and productivity of animals in four subtidal rocky reef habitats: The importance of small mobile invertebrates. *Marine Ecology Progress Series, 172*, 37–51.

Tegner, M. J., & Dayton, P. K. (2000). Ecosystem effects of fishing in kelp forest communities. *ICES Journal of Marine Science, 57*, 579–589.

Ulanowicz, R. (1986). *Growth and development: Ecosystems phenomenology*. New York: Springer.

Ulanowicz, R. (1997). *Ecology, the ascendent perspective. Complexity in ecological systems series*. New York: Columbia University Press.

Ulanowicz, R., & Puccia, C. (1990). Mixed trophic impacts in ecosystems. *Ceonoces, 5*, 7–16.

Uribe, R., Ortiz, M., Macaya, E. C., & Pacheco, A. S. (2015). Successional pattern of hard-bottom macrobenthic communities at kelp bed (*Lessonia trabeculata*) and barren ground sublittoral systems. *Journal of Experimental Marine Biology and Ecology, 472*, 180–188.

Valls, A., Coll, M., & Christensen, V. (2015). Keystone species: Toward an operational concept for marine biodiversity conservation. *Ecology Monographs, 85*, 29–47.

Vasas, V., Lancelot, C., Rousseau, V., & Jordán, F. (2007). Eutrophication and overfishing in temperate nearshore pelagic food webs: A network perspective. *Marine Ecology Progress Series, 336*, 1–14.

Vásquez, J. A. (2008). Production, use and fate of Chilean brown seaweeds: re-sources for a sustainable fishery. *Journal of Applied Phycology, 20*, 457–467.

Vásquez, J. A., & Buschmann, A. H. (1997). Herbivore-kelp interactions in Chilean subtidal communities: A review. *Revista Chilena de Historia Natural, 70*, 41–52.

Vásquez, J. A., Zúñiga, S., Tala, F., Piaget, N., Rodríguez, D. C., & Alonso Vega, J. M. (2013). Economic valuation of kelp forests in northern Chile: Values of goods and services of the ecosystem. *Journal of Applied Phycology, 26*, 1081. https://doi.org/10.1007/s10811-013-0173-6.

Villegas, M. J., Laudien, J., Sielfeld, W., & Arntz, W. E. (2007). *Macrocystis integrifolia* and *Lessonia trabeculata* (Laminariales; Phaeophyceae) kelp habitat structures and associated macrobenthic community off northern Chile. *Helgoland Marine Research, 62*, 33–43.

Walters, C., & Christensen, V. (2007). Adding realism to foraging arena predictions of trophic flow rates in Ecosim ecosystem models: Shared foraging arenas and bout feeding. *Ecological Modelling, 209*, 342–350.

Warwick, R. M., & Clarke, K. R. (1993). Comparing the severity of disturbance: A meta-analysis of marine macrobenthic community data. *Marine Ecology Progress Series, 92*, 221–231.

Wolff, M. (1994). A trophic model for Tongoy Bay – a system exposed to suspended scallop culture (Northern Chile). *Journal of Experimental Marine Biology and Ecology, 182*, 149–168.

Xavier, J. C., & Croxall, J. P. (2007). Predator-prey interactions: Why do larger albatrosses eat bigger squid? *Journal of Zoology, 271*, 408–417.

Exploring Harvest Strategies in a Benthic Habitat in the Humboldt Current System (Chile): A Study Case

Jorge E. González and Marco Ortiz

1 Introduction

The absence of long-term comparisons of the effects of fisheries on the ecological and economic sustainability of marine ecosystems is one of the most important gaps in assessing the consequences of different management and management strategies (Hundloe, 2000). The use of economic-productive criteria for the development of fishing activities is related to an increase in economic income, which should be compatible with the natural productivity of marine systems (Walters & Martell, 2004). Traditional research approaches with regard to fishery management have been oriented towards understanding productive processes under economic efficiency concepts. They have generally considered the dynamics of a single species (resource) using a reductionist approach. However, economic models based on this approach are deficient in addressing economic-productive processes because they are based on simplified assumptions about ecosystem functions and their responses to disturbances (Sanchirico & Wilen, 1999; Garcia & Cochrane, 2005). The development of unregulated or poorly planned extractive activities is usually accompanied by negative effects associated with changes in the functionality of natural

J. E. González (✉)
Departamento de Biología Marina, Facultad de Ciencias del Mar, Universidad Católica del Norte, Coquimbo, Chile
e-mail: jorge.gonzalez@ucn.cl

M. Ortiz
Laboratorio de Modelamiento de Sistemas Ecológicos Complejos (LAMSEC), Instituto Antofagasta, Universidad de Antofagasta, Antofagasta, Chile

Instituto de Ciencias Naturales Alexander von Humboldt, Facultad de Ciencias Naturales y Recursos Biológicos, Universidad de Antofagasta, Antofagasta, Chile

Departamento de Biología Marina, Facultad de Ciencias del Mar, Universidad Católica del Norte, Coquimbo, Chile

M. Ortiz, F. Jordán (eds.), *Marine Coastal Ecosystems Modelling and Conservation*, https://doi.org/10.1007/978-3-030-58211-1_6

systems. These deficits affect their productive capacity and, consequently, lead to inefficient economic performance (Beddington & Retnig, 1984; Lawson, 1984; Mullon et al., 2009).

The ecosystem approach focused on fisheries (EAF) is oriented in terms of ensuring socioeconomic and ecological objectives by maintaining ecosystem functions and facilitating the achievement of both objectives through effective management (Pikitch et al., 2004; Garcia & Cochrane, 2005; Livingston et al., 2005). In this framework, the ecological-economic balance of an ecosystem has been interpreted as part of a 'general equilibrium' of ecosystem sustainability (Templet, 1999; Teh et al., 2005; Mullon et al., 2009). Therefore, achieving economic-ecological sustainability objectives emerges from a commitment between both objectives. Success will ultimately depend on the *state of health of the ecosystem* (sensu Cheung & Sumaila, 2008).

The comparative assessments of performance between ecological and economic indicators of an ecosystem under exploitation correspond to a central aspect for understanding the consequences of fisheries at the ecosystem scope (Garcia & Cochrane, 2005). These associated indicators to evaluate the effect of different strategies of exploitation (or use of the fishing system) should consider the integration of profitability and effects on the ecosystem (Bonzon, 2000; Hundloe, 2000; Christensen & Walter, 2004; Ceriola et al., 2008). In the absence of historical indicators, dynamic modelling allows one to establish scenarios, using different mortality rates that permit the assessment of sustainability objectives for exploited ecosystems (Christensen & Walter, 2004). *Ecosim* corresponds to the dynamic module of the *Ecopath With Ecosim* programme (EwE; Christensen and Walter, 2004), which offers the possibility—through objective functions for management—to explore the impacts of fishery policies through formal optimization of economic, social and ecological objectives. Objective functions allow one to establish policies that maximize the economic income of the fishery/ecosystem and/or maximize the structure of the ecosystem. The former is based on the search for maximum profits, while the latter is oriented to maintain and/or maximize the structure of the ecosystem ('health'; Odum, 1969; Christensen, 1995). The ecosystem productivity of Tongoy Bay, Chile (Fig. 1a) is conditioned by the occurrence of periodic upwelling near the centre of the bay (Daneri et al., 2000). These upwelling dynamics have led to the development of important benthic fisheries and consequent human interventions (Wolff & Alarcón, 1993; Wolff, 1994; Ortiz & Wolff, 2002). The total landings of the benthic resources from Tongoy Bay have fluctuated substantially since 1985, reaching a peak value in 1992 of ~300 tons. The main exploited resources are predatory crabs, such as *Romaleon polyodon*, the scallop *Argopecten purpuratus* and clams. The two last resources are prey of *R. polyodon*.

Since 2012, landings of resources have shown a downward trend accompanied by changes in the composition of exploited species. Further, the macroscopic properties determined for Tongoy Bay have demonstrated an increase in the state of health compared to past conditions, a phenomenon that could explained by a reduced fishing pressure on this benthic ecosystem in recent years as a consequence of the establishment of Territorial User Rights for Fishing (TURF) management strategies

(González et al., 2016). The low fishing pressure has positively impacted exploited species, as well as the structure and functioning ('health') of the ecosystem (González et al., 2016). Therefore, we will adopt a dynamic trophic network analysis for assessing the ecological/economic performance of the exploited benthic ecosystem of Tongoy Bay under a TURF regimen, using a multispecies harvest strategy.

2 Geographical Area and Modelling

Tongoy Bay (Fig. 1a) is located in north-central Chile (30°12′S 71°34′W). This bay has high productivity due to the presence of a seasonal (spring and summer) upwelling (Fonseca & Farias, 1987). Seasonal upwelling produces high phytoplankton biomass, which in turn supports fishing and scallop (*A. purpuratus*) aquaculture (Boré et al., 1993). Natural stocks of this scallop and *Chondracanthus chamissoi* alga were depleted, and benthic landings experienced a remarkable reduction until 1996 (González et al., 1996; Stotz & Aburto, 2013). In 1998, management areas for benthic resource exploitation were established at Tongoy Bay, under a TURF, as a measure to reduce fishing pressure. This regimen has allowed recovery of the system, but with lower harvest compared to before the 1990s (Fig. 1b).

2.1 *Baseline Information for Models*

The analysis was carried out by considering three constructed stationary trophic models of the benthic ecosystem of Tongoy Bay (Fig. 2), used the EwE programme (Christensen 1995). These corresponded to the years 1992, 2002 and 2012, based on Wolff (1994), Ortiz & Wolff (2002) and González et al. (2016). For more details

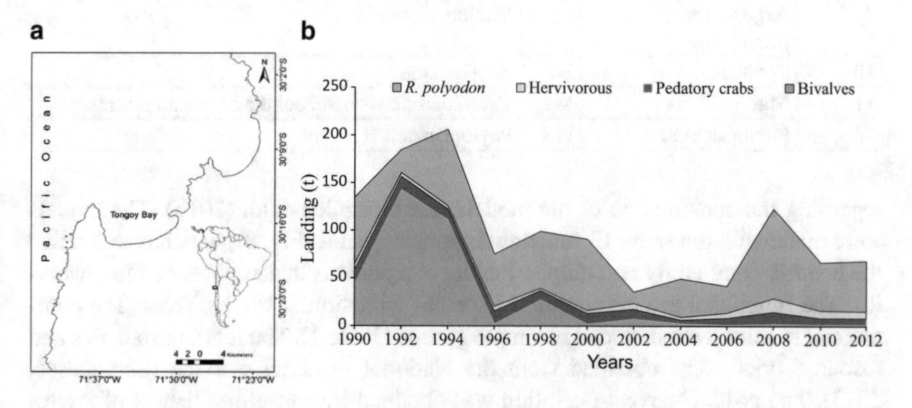

Fig. 1 Study area (**a**) and landing of benthic resource (**b**) for Tongoy bay benthic ecosystem

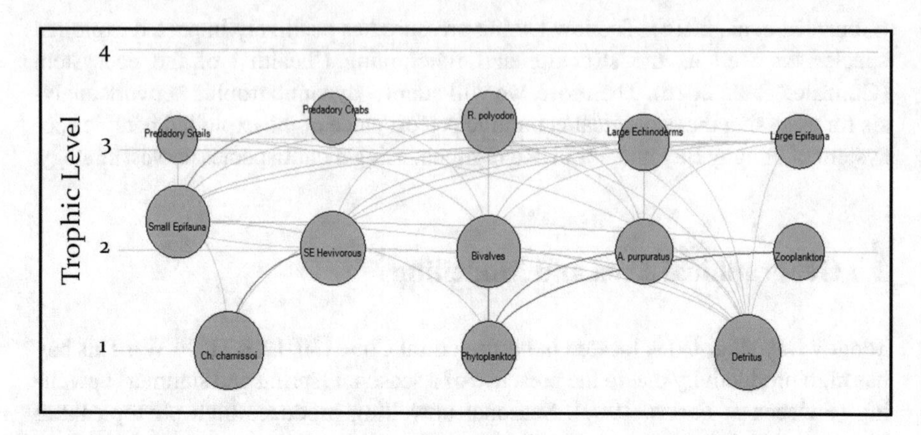

Fig. 2 Trophic model for the benthic ecosystem of Tongoy bay. The model represents the year 2012. Vertical position approximates the trophic level. The circle size is proportional to the compartment (species and/or functional groups) biomass (g wet weight m^{-2}). The numbers inside circles correspond to name of species or functional groups (see Table 1 for details)

Table 1 Species and functional groups considered in the trophic model. The commercial species (in bold) are included as individual compartments in the model

Number	Trophic group name	Code	Conspicuous spp.
1	Predatory snails	PS	Xanthochorus sp. and Priene sp.
2	Predatory crabs	PC	Cancer sp. and Homalaspis plana
3	**Romalion polyodon**	RP	Predator crab
4	Seastars	SS	Heliaster helianthus, Meyenaster gelatinosus and Luidia magellanica)
5	Small epifauna herbivores	SEH	Tegula sp. and Fissurella spp.
6	Large epifauna	LE	Pagurus sp. and Alpheus sp.
7	Small epifauna	SE	Caprella sp. and Nereis sp.
8	Bivalves	Biv	Mulinia edulis and Ensis macha
9	**Argopecten purpuratus**		Scallop
10	Zooplankton	Zoo	Zooplankton
11	**Macrophytes**	Ma	Chondracanthus chamissoi the dominant macrophyta
12	Phytoplankton	Phy	Phytoplankton

regarding the construction of the models, see González et al. (2016). The models were made with the same 13 functional groups/species (Fig. 2), which characterize the benthic community and emphasize the components that support benthic fisheries. The functional groups comprised species with similar trophic roles. The commercial species were included in simple groups (Table 1). The landing statistics and resource price were obtained from the National Fisheries Service (Sernapesca, 2013). The cost of harvest operation was obtained by consulting fishers of Puerto Aldea cove, which is adjacent to the study area.

2.2 *Simulation of Harvest by Changes in Fishing Mortality*

In order to simulate the effect of fishery rate over exploited resources and the structure ecosystem, the *Ecosim* module (Walters et al., 1997) was used. *Ecosim* employs the set of linear equations utilized in *Ecopath* but re-expresses them as differential equations. The basic *Ecosim* equation is represented as:

$$\frac{dB_i}{d_t} = f(B) - M_o B_i - F_i - B\sum_{j=1}^{n} c_{iji}\left(B_{i,}B_j\right), \tag{1}$$

where *B* is the biomass; *Mo* is the mortality rate not generated by fishing or predation; *Fi* is the fishing mortality coefficient; and *f*(*B*) represents the production function if the group is a primary producer or a growth function if the group is a consumer. Moreover, *cij* (*Bi*, *Bj*) is the function to predict the consumption of prey *i* by predator *j*. The simulations respond to changes in fishing mortality for the component under harvest condition, according to the observed in the stationary models.

2.3 *Analysis of Optimization of Economic and Ecological Scenarios*

Optimization analyses were performed through the selection of ecological and/or economic maximization objectives. The task was to identify a single performance indicator based on the overall performance of a combined value of fishing operations over the resource pool. The modification of fleet fishing efforts (fishing mortality) allowed maximizing indicators associated with economic and/or ecological objectives (Christensen & Walters, 2004).

The economic objective is based on maximizing net benefits (NB) of fishing as result of catch value (Value = catch level x price of each resource) minus the cost of fishing (fixed costs + variable costs). The fleet definition (Table 2) was set as the baseline scenario and used for the discount rates (to discount the value of future catches relative to the present value). A traditional discount rate (4%) as well as a 10% intergenerational discount rate were considered (Ainsworth & Sumaila, 2005). The intergenerational discount rate introduces the net present value (NPV), estimating the benefits received by the current generation (calculated at a standard discount rate) plus the value of the benefits received by future generations (Sumaila & Walters, 2005).

The flow of benefits derived from fishing, under a conventional discount rate, was expressed as net present value (NPV) as follows:

$$\text{VPN} = \sum_{t=0}^{T}\left(d^t * NB_t\right), \tag{2}$$

Table 2 Reference of fishing mortality ($F_{Ecopath}$) and catch for benthic resources inhabit Tongoy bay for 1992, 2002 and 2012 models. The inputs for economic estimates of unit fleet are included

	F_Ecopath (año-1)			Price ($/gr)
Groups- species/models	1992	2002	2012	1992-2012
Predadory Snails	0.18	0.25	0.03	1.50
R. Polyodon	0.21	0.04	0.09	1.00
Bivalves	0.37	0.29	0.00	1.20
Purpuratus	0.58	0.33	0.00	4.00
Ch. chamissoi	–	–	0.01	0.60
Definition of fleet				
Fixe cost (%)				0
Operational cost (%)				30
Net profit (%)				70
Discount rate (%)				4
Intergeneration discount rate (%)				10

where *NB* is the net profit by year *t*, and *d* is the discount factor, given by

$$d = \frac{1}{(1+\delta)}, \tag{3}$$

where δ is the discount rate.

The ecological objective considers the maximization of the structure of the ecosystem ('health'). This approach is based on Odum (1969), who describes that 'mature' ecosystems are dominated by large and long-lived organisms. As a measure of the structure of the ecosystem, the ratio of biomass/specific production of each resource (as measure of the longevity of each group) was used. These values correspond to the inverse of the P/B ratio:

$$\text{Ecological index} = \sum_{i=1}^{n} \frac{B_i}{P_i}, \tag{4}$$

where B_i and P_i are the biomass and production of functional group *i*, respectively.

The search for economic/ecological sustainability criteria was based on the routine 'fisheries policy searches' that is incorporated in the *Ecosim* module (Christensen & Walters, 2004). The parameters per year of the fleet were defined (Table 2) considering a unitary fleet (sole owner) and a fishery rate (fishing mortality [F]) obtained from the stationary models (*Ecopath*). For each model, a projection of 20 years was established using a mixed flow control (Zetina-Rejón et al., 2004; Ainsworth & Sumaila, 2005).

The objective function for the economic/ecological criteria was established as a 'multi-criterion objective', which is represented as the weighted sum of the economic and ecological criteria as follows: alternative weights of 1:5, 5:1 and 3:3 were established for economic and ecological objectives. The first two scenarios were used to counteract optimization conflicts, while the third was employed to evaluate symmetrical compensation between both objectives. The objective function is calculated from the weighted sum (w) of the indices that represent the objectives of the policies economic and ecological, according to:

$$\text{Funtion}_{\text{total}} = \omega_{\text{economic}} * \text{NPV} + \omega_{\text{ecosystem}} * \text{Ecological index}, \quad (5)$$

where NPV = net present value (Eq. (1)) and the ecological index estimated according to Eq. (2) and Ecological index by Eq. (4).

The optimization is based on a Davidon–Fletcher–Powell (DFP) nonlinear procedure, which involves testing alternative magnitudes for fishing mortality in order to improve the objective function. The variation scheme of the parameters used by DFP corresponds to the 'conjugate gradient' method, which generates tests of alternative values of the parameters (mortality) to approach the objective function (Christensen and Walter, 2004).

2.4 *Ecosystem Maturity Indicator*

To evaluate the degree of development of the system under different scenarios, the *Ascendency* theoretical framework (sensu Ulanowicz, 1986) was used. It is a network analysis that allows one to evaluate the degree of growth (activity) and flow coherence (organization) of an ecosystem. *Ascendency* describes the trend of an ecosystem after natural or anthropogenic disturbances (Costanza & Mageau, 1999; Walters & Martell, 2004); it permits one to compare ecosystem trajectories at different times (Ulanowicz, 1997; Heymans, 2003; Christensen et al., 2005). Therefore, different macroscopic indices associated with *Ascendency* were estimated for each scenario under 20 years of simulation. For more details about *Ascendency*, see Chap. 5 in this book.

3 Modelling Outcomes

Table 3 presents the obtained outcomes using the maximization procedure based on economical, ecological and the combination of both functions. For the three models (years), the optimization showed an opposite tendency in relation to the indicators based on the economic objectives and ecosystem structure. The increase in weight of the economic objective over the ecosystem (5:1) led to an obvious increase in the economic value associated with an increase in the fleet and catches. These scenarios exhibited a decrease in the ecological value for the years 1992 and 2002 (Fig. 3). By contrast, for the year 2012 under the economic optimization objective, the highest values were achieved for all indices—with a 6.2-fold increase in economic value—while maintaining the ecological index (Fig. 3). On the other hand, an increase of

Table 3 Changes in the performance of economic, ecological and yield indexes of Tongoy bay for the years 1992, 2002 and 2012, based on different weights on optimization policies: economic, ecosystem structure (ecological) and economic-ecological combination (Econ-Ecol)

	Economic			Ecological			Econ-Ecol		
Indexes/models	1992	2002	2012	1992	2002	2012	1992	2002	2012
Economic value (end/start)	1.4	2.3	6.2	0.0	0.9	5.1	1.4	2.3	5.6
Ecological value (end/start)	0.9	0.8	1.1	1.16	1.2	1.4	0.9	1.0	1.1
Effort (end/start)	1.9	3.8	12.1	0.0	0.9	9.1	1.8	3.6	10.1
Harvest (end/start)	1.5	2.6	5.8	0.02	0.9	5.6	1.5	2.6	5.8

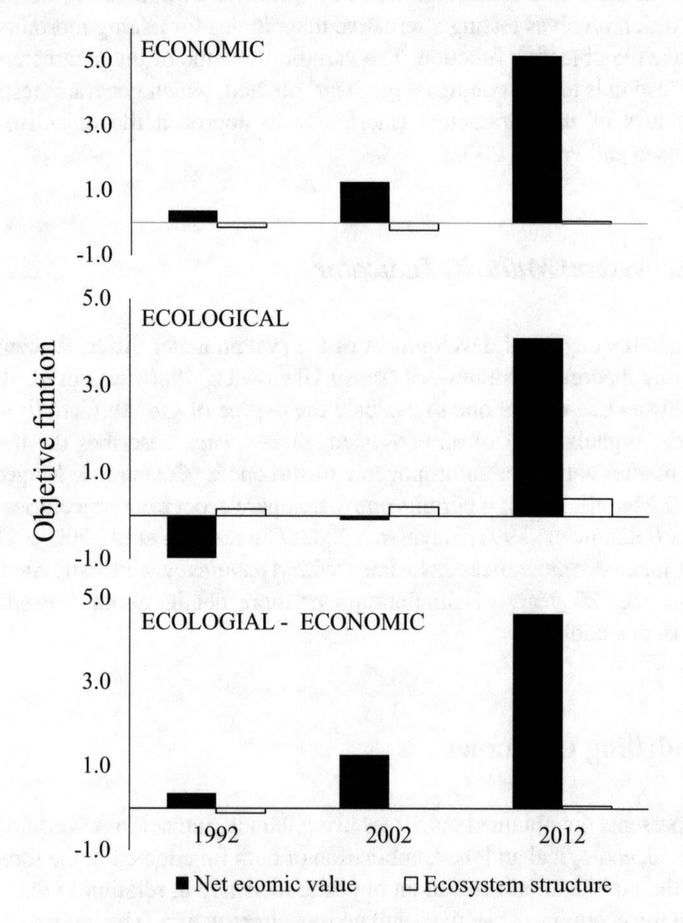

Fig. 3 Changes in economic value and ecological 'stability' index (measured as ecological structure) for the optimization of economic, ecological and combined objectives of Tongoy bay for the years 1992, 2002 and 2012

the relative weight towards the ecosystem structure over the economic (5:1) showed an increase in ecological value, but a reduction in the effort, catches and consequent economic performance for the years 1992 and 2002. Nevertheless, the year 2012 exhibited a weak decrease in economic value compared to the same index under the scenario of economic optimization for the years 1992 and 2002. The scenario that set equal weight for the economic and ecological criteria generated intermediate responses in the both indices. Under this scenario, only the year 1992 showed a reduction in the ecological index (Table 3).

The baseline fish mortality levels ($F_{Ecopath}$; year^{-1}) obtained for the three optimization criteria are shown in Fig. 4. These values represent relative changes in mortality for each criterion after optimization with regard to the baseline situation (stationary models). In the economic and economic-ecological scenarios, fishing mortality increased from 1992 to 2012, with the highest magnitudes for the year 2012. In the ecological optimization scenario, fishing mortality was minimized to reduce the effects on ecosystem structure. For 1992, fishing mortality was practically reduced to zero, slightly increased during the year 2002 and remarkably increased in 2012. In the multi-criteria scenario, the pattern of mortality increased in a similar fashion compared to the economic optimization model (Fig. 4).

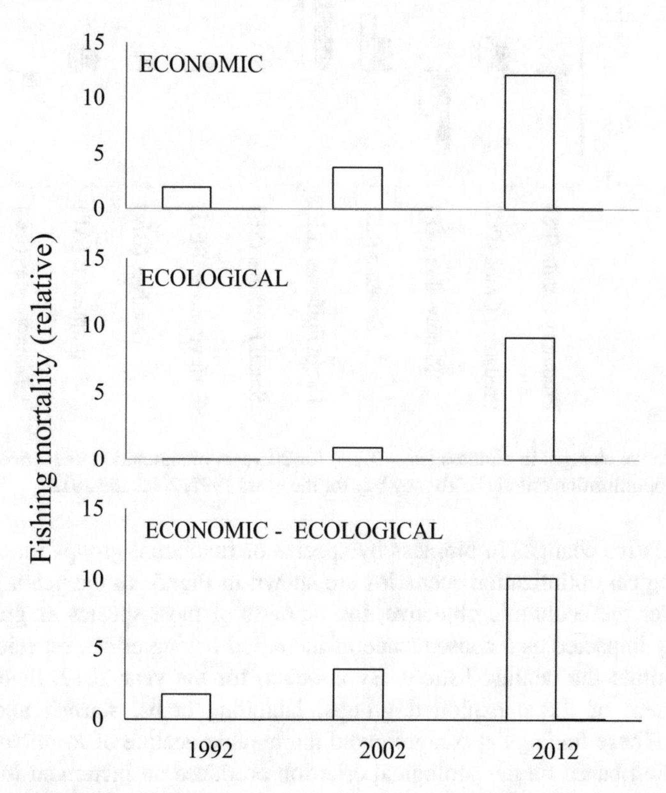

Fig. 4 Changes in relative fishing mortality ($F_{final}/F_{inicial}$) under alternative optimization criteria for Tongoy bay for the years 1992, 2002 and 2012

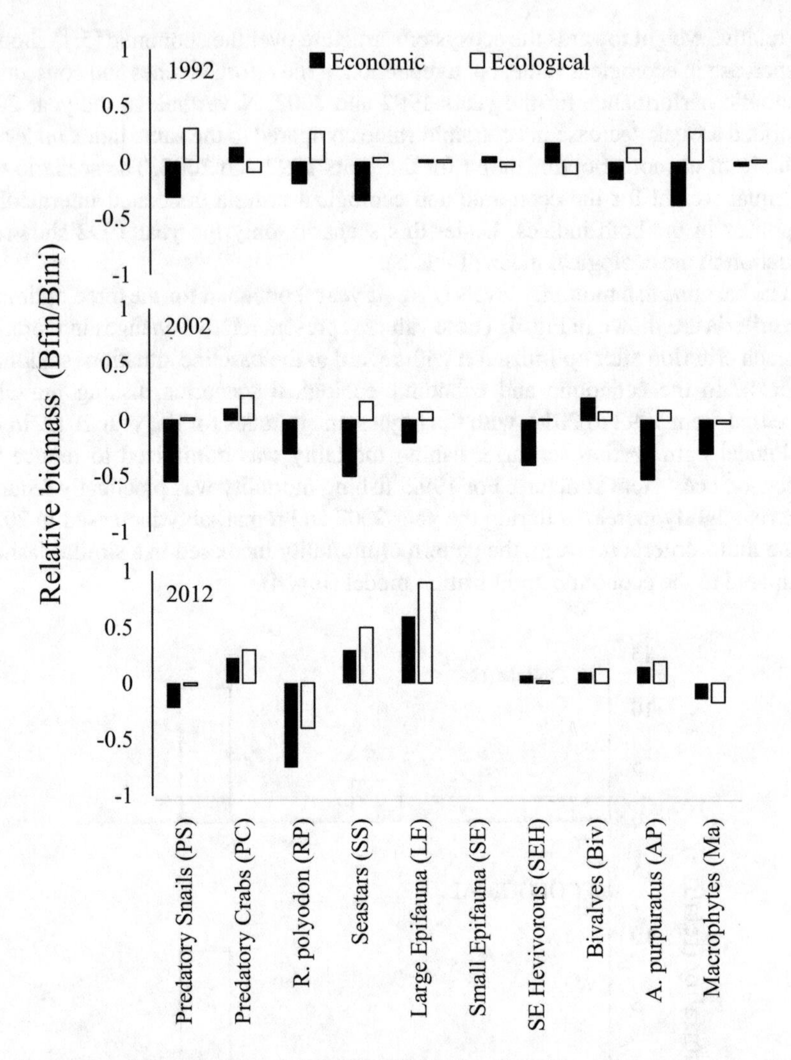

Fig. 5 Relative changes in biomass ($B_{\text{final}}/B_{\text{inicial}}$) for 20 years of simulation under economic and ecological optimization criteria of Tongoy bay for the years 1992, 2002 and 2012

The relative changes in biomass by species or functional groups for economic and ecological optimization scenarios are shown in Fig. 5. In the years 1992 and 2002 under the economic objective, the biomass of most species or groups was negatively impacted as a consequence of increased fishing effort, especially those that constitute the benthic fishery. By contrast, for the year 2012, there was an enhancement of the unexploited groups, including crabs, starfish and macro-epifauna. These findings also highlighted the notable decline of *R. polyodon*. The optimization based on the ecological criterion produced an increment in the final biomass in the most of groups, especially in those located at high trophic levels. For the year 2012, after the economic criterion, there was a remarkable increase in the biomass of unexploited functional groups, such as sea stars, crabs and epifauna. On the other hand, the crab *R. polyodon* presented a notable biomass reduction under the ecological and the economic objective. The macroscopic properties of *Ascendency* and *Overhead* for the three models after optimizations are shown in Fig. 6. The models for the years 2002 and 2012 achieved the highest values for *Ascendency* and *Overhead*—increasing the *Capacity* of the system—in comparison to the magnitudes observed for the model of year 1992.

Management of exploited ecosystems is a main challenge for achieving economic and ecological sustainability (Garcia & Cochrane, 2005; Cheung & Sumaila, 2008). The marine ecosystem of Tongoy Bay was heavily exploited until the early 1990s; this level subsequently decreased as a result of the spatial control of exploitation levels under the TURF management strategy (González et al., 2006). This tendency is demonstrated by decreasing the levels of landings observed during the last 20 years. Long-term optimizations for Tongoy ecosystem have been applied as a comparative strategy to expose the changes in the 'stability' of the benthic ecosystem. In this regard, the model for the year 2002 exhibited the greatest system capacity to resist—under ecological and economic optimization criteria—high levels of fishing disturbance. This finding agrees with the concept that a more 'stable' system would present greater margins for economic optimization without negatively affecting ecological 'stability', and vice versa (Teh et al., 2005). It is relevant to note that the best balance performance between economic and ecological criteria for optimization is achieved in the model representing the most recent and less disturbed condition (year 2012).

When the ecological criteria were optimized, the ecosystem properties were improved at the expense of economic performance (Bundy, 2002; Vasconcellos &

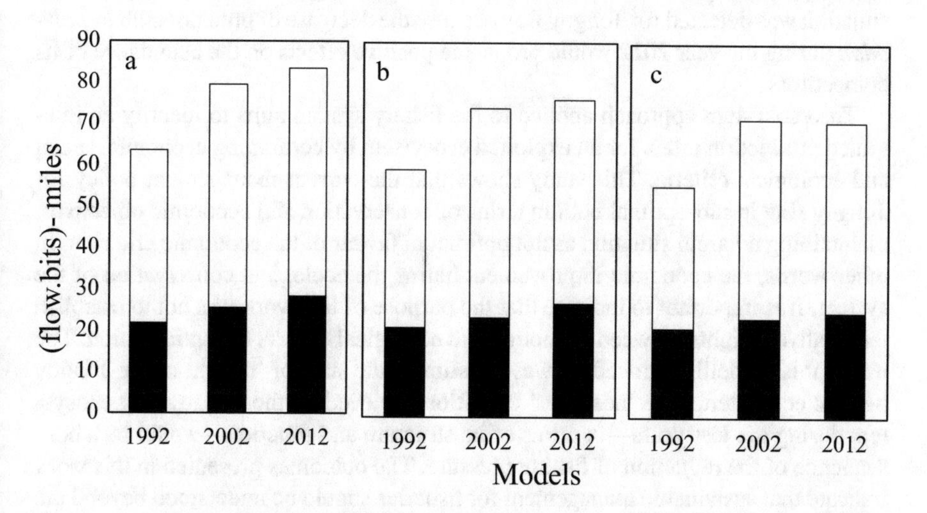

Fig. 6 Macroscopic indices: *Ascendency* (black) and *Overhead* (white) for the Tongoy bay ecosystem in the 1992, 2002 and 2012 years simulating under ecological (**a**), ecological-economic (**b**) and economic (**c**) optimization criteria

Heymans, 2002; Zetina-Rajón et al., 2004). Only the model for the year 2012 allowed optimizing the economic objective without 'destabilizing' impacts in the ecological system. This outcome could indicate that in 2012 the ecosystem was healthier than before. The economic optimization for Tongoy Bay observed for the year 2012, although desired, would not necessarily be appropriate. Indeed, the increase in fishing intensity is associated with the drop of abundance of commercial species as well as other species (Arreguín-Sánchez et al., 2004; Zetina-Rejón et al., 2004). An intermediate reduction of the fishing effort would improve the 'stability' of exploited ecosystems (Walters & Martell, 2004). However, Zetina-Rejón et al. (2001) suggested that highly exploited ecosystems are difficult to improve in terms of fishery production (Ortiz & Levins, 2017).

4 Conclusions

The ecological criterion for optimization seeks to reduce the negative impacts on ecological 'stability', as was determined for the year 1992. Given that the lowest 'health' condition of the ecosystem was estimated for 1992, its recovery could have been achieved through decreasing fishing mortality with the consequent decline of economic benefits. The economic criterion for optimization is based on the most profitable exploitation of resources at the expense of competitors and predators (Christensen & Walter, 2004). In this sense, Vasconcellos & Heymans (2002) discovered that predators of higher trophic levels are negatively impacted when the economic criterion is optimized. Likewise, Hooper et al. (2005) indicated that a drop in predators would lead to a reduction in the ecological 'stability'. A particular situation was detected for Tongoy Bay because the decrease of predator crab *R. polyodon* during the year 2012 would propagate positive effects on the abundance of its competitors.

The ecosystem approach applied to the fishery system aims to identify an integrated production index for an exploited ecosystem by combining economic, social and ecological criteria. This study shows that the current management policy for Tongoy Bay is sub-optimal both in terms of conservation and economic objectives, maintaining a Pareto situation as not optimal in favour of the economic criterion. In other words, the economic improvement harms the ecological conservation of the system. It is important to indicate that the purpose of this work was not to establish the relative weights between economic and ecological criteria for optimization. The use of this modelling procedure was to estimate the state of 'health' of the Tongoy benthic ecosystem. The 'healthier' condition achieved by the Tongoy Bay ecosystem during the last years—in terms of its structure and function—would be a consequence of the reduction of fishing pressure. The outcomes presented in this work indicate that sustainable management for fisheries should be understood beyond the single-resource classical approach. Indeed, it must be recognized that ecosystems generate direct and indirect goods and services related to their particular ecosystem properties (Walters & Martell, 2004). In practical terms, an effective management for Tongoy Bay should consider a re-definition of artisanal fisheries management co-varying with natural ecosystem changes (Walters & Parma, 1996; Ortiz & Levins, 2011, 2017).

Acknowledgement This research was supported from Comisión Nacional de Investigación Científica y Tecnológica (CONICYT). J.G. thanks Fondecyt #3170914 Postdoctoral Grant.

References

Aburto, J., & Stotz, W. (2013). Learning about TURFs and natural variability: failure of surf clam management in Chile. *Ocean & Coastal Management, 71*, 88–98.

Ainsworth, C. H., & Sumaila, U. R. (2005). Intergenerational valuation of fisheries resources can justify long-term conservation: A case study in Atlantic cod (*Gadus morhua*). *Canadian Journal of Fisheries and Aquatic Sciences, 62*, 1104–1110.

Arreguín-Sánchez, F., Zetina-Rejón, M., Manickchand-Heileman, S., Ramırez-Rodrıguez, M., & Vidal, L. (2004). Simulated response to harvesting strategies in an exploited ecosystem in the southwestern Gulf of Mexico. *Ecological Modelling, 172*(2), 421–432.

Beddington, J. R., & Reting, R. B. (1984). Criterios para la ordenación del esfuerzo de pesca. *FAO Document Technique sur les Pêches, 243*, 44.

Bonzon, A. (2000). Development of economic and social indicators for the management of Mediterranean Fisheries. *Marine and Freshwater Research, 51*, 493–500.

Boré, D., Blanco J.L., Acuña E., Moraga J., Olivares J., Mujica A., & Uribe E. (1993). Evaluación de la distribución de recursos pelágicos de la IV Región y condiciones oceanográficas asociadas. *Informe Técnico IFOP, Chile. Proyecto F.I.P.* 57 pp

Bundy, A. (2002). Exploring multispecies harvesting strategies on the eastern Scotian shelf with Ecosim. In: Pitcher, T., Cochrane K. (Eds.), The use of ecosystem models to investigate multispecies management strategies for capture fisheries. *Fisheries Centre Research Reports, 10*(2), 112–117.

Ceriola, L., Accadia, P., Mannini, P., Massa, F., Milone, N., & Ungaro, N. (2008). A bio-economic indicators suite for the appraisal of the demersal trawl fishery in the Southern Adriatic Sea (Central Mediterranean). *Fisheries Research, 92*, 255–267.

Cheung, W. W., & Sumaila, U. R. (2008). Trade-offs between conservation and socio-economic objectives in managing a tropical marine ecosystem. *Ecological Economics, 66*(1), 193–210.

Christensen, V. (1995). Ecosystem maturity—towards quantification. *Ecological Modelling, 77*(1), 3–32.

Christensen, V., & Walter, C. (2004). Ecopath with ecosim: Methods, capabilities and limitations. *Ecological Modelling, 172*, 109–139.

Christensen, V., Walters, C. J., & Pauly, D. (2005). Ecopath with Ecosim: a user's guide. *Fisheries Centre, University of British Columbia, Vancouver*, 154.

Costanza, R., & Mageau, M. (1999). What is a healthy ecosystem? *Aquatic Ecology, 33*, 105–115.

Daneri, G., Dellarossa, V., Quiñones, R., Jacob, B., Montero, P., & Ulloa, O. (2000). Primary production and community respiration in the Humboldt Current System off Chile and associated oceanic areas. *Marine Ecology Progress Series, 197*, 41–49.

Fonseca, T., & Farías, M. (1987). Estudio del proceso de surgencia en la costa chilena utilizando percepción remota. Investigaciones *Pesqueras, 34*, 33–46.

Garcia, S., & Cochrane, K. (2005). Ecosystem approach to fisheries: A review of implementation guidelines. *ICES Journal of Marine Science, 62*, 311–318.

González, J., & Meneses, I. (1996). Differences in the early stages of development of gametophytes and tetrasporophytes of Chondracanthus chamissoi (C. Ag.) Kützing from Puerto Aldea, northern Chile. *Aquaculture, 143*(1), 91–107.

González, J., Ortiz, M., Rodríguez-Zaragoza, F., & Ulanowicz, R. E. (2016). Assessment of long-term changes of ecosystem indexes in Tongoy Bay (SE Pacific coast): Based on trophic network analysis. *Ecological Indicators, 69*, 390–399.

González, J., Stotz, W., Garrido, J., Orensanz, J. M., Parma, A. M., Tapia, C., & Zuleta, A. (2006). The Chilean TURF system: How is it performing in the case of the loco fishery? *Bulletin of Marine Science, 78*(3), 499–527.

Heymans, J. (2003). Revised models for Newfoundland for the time periods 1985-87 and 1995-97. In Ecosystem Models of Newfoundland and Southeastern Labrador: Additional information and analyses for 'Back to the Future'. Edited by Johanna J. Heymans. *Fisheries Centre Research Reports. 11*(5), 79 pp.

Hooper, D. U., Chapin, F. S., Ewel, J. J., Hector, A., Inchausti, P., Lavorel, S., Lawton, H., et al. (2005). Effects of biodiversity on ecosystem functioning: a consensus of current knowledge. *Ecological Monographs, 75*, 3–35.

Hundloe, T. J. (2000). Economic performance indicators for fisheries. *Mar. Freshw. Res., 51*, 485–491.

Lawson, R. M. (1984). *Economics of fisheries management*. New York: Praeger Publishers.

Livingston, P. A., Aydin, K., Boldt, J., Ianelli, J., & Jurado-Molina, J. (2005). A framework for ecosystem impacts assessment using an indicator approach. *ICES Journal of Marine Science, 62*, 592–597.

Mullon, C., Shin, Y., & Cury, P. (2009). NEATS: a network economics approach to trophic systems. *Ecological Modelling, 220*(21), 3033–3045.

Odum, E. P. (1969). The strategy of ecosystem development. *Sustainability: Sustainability, 164*, 58.

Ortiz, M., & Wolff, M. (2002). Trophic models of four benthic communities in Tongoy Bay (Chile): comparative analysis and assessment of management strategies. *Journal of Experimental Marine Biology and Ecology, 268*(2), 205–235.

Ortiz, M., & Levins, R. (2011). Re-stocking practices and illegal fishing in northern Chile (SE Pacific coast): a study case. *Oikos, 120*(9), 1402–1412.

Ortiz, M., & Levins, R. (2017). Self-feedbacks determine the sustainability of human interventions in eco-social complex systems: Impacts on biodiversity and ecosystem health. *PloS one, 12*(4), e0176163.

Pikitch, E., Santora, C., Babcock, E., Bakum, A., Bonfil, R., Conover, D., Dayton, P., Doukakis, P., Fluharty, D., Heneman, B., Houde, E., Link, J., Livingston, P., Mangel, M., McAllister, M., Pope, J., & Sainsbury, K. (2004). Ecosystem-based fishery management. *Science, 305*, 346–347.

Sanchirico, J. N., & Wilen, J. E. (1999). Bioeconomics of spatial exploitation in a patchy environment. *Journal of Environmental Economics and Management, 37*, 129–150.

Sernapesca, (2013). Anuario Estadístico de Pesca. *Ministerio de Economía Fomento y Reconstrucción*, República de Chile. 115 pp.

Sumaila, U. R., & Walters, C. J. (2005). Intergenerational discounting: A new intuitive approach. *Ecological Economics, 52*, 135–142.

Teh, L., Cabanban, A. S., & Sumaila, U. R. (2005). The reef fisheries of Pulau Banggi, Sabah: A preliminary profile and assessment of ecological and socio-economic sustainability. *Fisheries Research, 76*(3), 359–367.

Templet, P. H. (1999). Energy, diversity and development in economic systems; An empirical analysis. *Ecological Economics, 30*(2), 223–233.

Ulanowicz, R.E. (1986). Growth and Development: Ecosystems Phenomenology. Springer, NY, 203 pp.

Ulanowicz, R. E. (1997). *Ecology, the Ascendent perspective. Complexity in Ecological Systems Series* (p. 201). NY: Columbia University Press.

Vasconcellos, M., & Heymans, J. A. B. (2002). The use of Ecosim to investigate multispecies harvesting strategies for capture fisheries of the Newfoundland-Labrador Shelf. In: Pitcher, T.,

Cochrane, K. (Eds.), The Use of Ecosystem Models to Investigate Multispecies Management Strategies for Capture Fisheries. *Fish. Centre Res. Rep., 10*(2), 68–72.

Walters, C., & Martell, S. J. D. (2004). *Fisheries ecology and management* (p. 399). Princeton: Princeton University Press.

Walters, C., & Parma, A. M. (1996). Fixed exploitation rate strategies for coping with effects of climate change. *Canadian Journal of Fisheries and Aquatic Sciences, 53*(1), 148–158.

Walters, C. J., Christensen, V., & Pauly, D. (1997). Structuring dynamic models of exploited ecosystems from trophic mass balance assessments. *Rev. Fish Biol. Fish., 7*, 139–172.

Wolff, M. (1994). A trophic model for Tongoy Bay –a system exposed to suspended scallop culture (Northern Chile). *Journal of Experimental Marine Biology and Ecology, 182*, 149–168.

Wolff, M., & Alarcón, E. (1993). Structure of a scallop Argopecten purpuratus (Lamarck, 1819) dominated subtidal macro-invertebrate assemblage in northern Chile. *Journal of shellfish Research, 12*(2), 295–304

Zetina-Rejón, M., Arreguín-Sánchez, F., & Chávez, E. A. (2001). Using and ecosystem modeling approach to assess the management of a Mexican coastal lagoon system. *CalCOFI Reports, 42*, 88–96.

Zetina-Rejón, M. J., Arreguı́, F., & Chávez, E. A. (2004). Exploration of harvesting strategies for the management of a Mexican coastal lagoon fishery. *Ecological Modelling, 172*(2), 361–372.

How Much Biomass Must Remain in the Sea After Fishing to Preserve Ecosystem Functioning? The Case of the Sardine Fishery in the Gulf of California, Mexico

Francisco Arreguín-Sánchez, Pablo del Monte-Luna, Mirtha O. Albañez-Lucero, Manuel J. Zetina-Rejón, Arturo Tripp-Quezada, T. Mónica Ruiz-Barreiro, and Juan C. Hernández-Padilla

1 Introduction

It is currently well recognized that many fishery resources are being exploited at levels close to the limit of their production capacity; in some cases, this limit has been exceeded (Murawski, 2010; Froese et al., 2011; Mansfield, 2011; Palkovacs, 2011; Pikitch, 2012; Worm & Branch, 2012; Colloca et al., 2013). The concerns about these conditions are related to the negative consequences for the sustainability of fishing systems as sources of food and livelihoods for the human population. One issue under discussion is industrial fishing, particularly on small pelagic fishes. It is assumed that the capture of large amounts of biomass can affect the dynamic balance of ecosystems and cause ecosystems to deteriorate (Smith et al., 2011; García, 2011; Garcia et al., 2012; Roux et al., 2013; Froese et al., 2016; Tommasi et al., 2017). Within this discussion, the industrial fisheries of small pelagic fish stand out. These fish species are herbivores or zooplankton consumers with short longevity; because of their abundance and position in the trophic pyramid, they are considered foraging species within the ecosystem (Cury et al., 2000; Bakun et al., 2009; Rosa et al., 2010; Griffiths et al., 2013; Hilborn et al., 2017). The fundamental concern from the point of view of the fishery resource is to avoid overfishing. However, in terms of sustainability, the scientific question is how much biomass can be removed without affecting the ecological role of the species? The concern is accentuated

F. Arreguín-Sánchez (✉) · P. del Monte-Luna · M. O. Albañez-Lucero · M. J. Zetina-Rejón
A. Tripp-Quezada · T. M. Ruiz-Barreiro
Instituto Politécnico Nacional, Centro Interdisciplinario de Ciencias Marinas, La Paz, México
e-mail: farregui@ipn.mx

J. C. Hernández-Padilla
Instituto Politécnico Nacional, Centro Interdisciplinario de Ciencias Marinas, La Paz, México

Instituto Nacional de Acuacultura y Pesca, Centro Regional de Investigación Pesquera de La Paz, La Paz, México

M. Ortiz, F. Jordán (eds.), *Marine Coastal Ecosystems Modelling and Conservation*, https://doi.org/10.1007/978-3-030-58211-1_7

because the populations of these fish are also vulnerable to environmental variations, and variations in their population can reach the order of thousands or millions of tons per year (Grbec et al., 2002; Chavez et al., 2003; Tourre et al., 2007; Fréon et al., 2008). This situation represents an additional problem for the management of small pelagic fish resources beyond the scientific question asked above; that problem has to do with the ability of management to deal with the natural variability in pelagic fish abundance and availability.

Recognizing the effects of medium- and long-term environmental trends on populations, it is recognized that populations respond according to the life history traits of their species but also according to their interdependencies with other species. Both effects propagate throughout the trophic network. As a result, populations respond and adapt at the individual level, and ecosystems are reconfigured. If changes in the environment that affect populations, such as climate change, show medium- to long-term trends, it is to be expected that populations will change continuously and that ecosystems are also continuously reconfigured. Fishing activity is inserted into these patterns, removing biomass and affecting the populations and the ecosystem (i.e. Grbec et al., 2002; Tourre et al., 2007; Bakun et al., 2009; Laugen et al., 2014).

In this context, and for management purposes, the concept of sustainability comes into play as the basis of public policy for management. In principle, sustainability refers to the long-term maintenance of levels of exploitation of the production capacity of the populations in an ecosystem. At present, the concept of the sustainability of the resources of an ecosystem that is experiencing continuous reconfiguration is one of the great challenges of science in support of management. In relation to the small pelagic fishes and the role they play as the base of the trophic pyramid, the scientific question to answer is how much biomass must remain in the sea after fishing to sustain the trophic functions of the ecosystem and to ensure sustainability?

To answer this question, and to illustrate a way to address this problem, the sardine fishery of the central Gulf of California, Mexico, was chosen as a case study. Seven small pelagic species are captured in the Gulf of California, and they are mainly used for fishmeal. The total catch volumes of these species are highly fluctuating, varying in the last four decades between 100 and 650 thousand tons per year (Fig. 1), in which the various species take part in varying proportions over time (Fig. 2).

1.1 The Trophic Network of the Central Gulf of California

The trophic model of the ecosystem reported by Arreguín-Sánchez et al. (2002), based on the *Ecopath* model (Christensen & Pauly, 1992), was modified to incorporate the seven species of small pelagic fishes that make up commercial catches in the Gulf of California in a disaggregated way. The model is composed of 37 functional groups (Annex): 21 fish, 5 macrocrustaceans, 2 molluscs and the following groups:

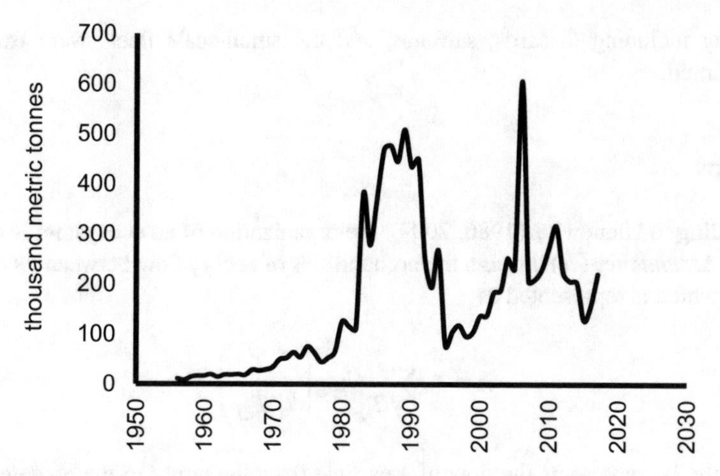

Fig. 1 Historical catches of small pelagic fishes in the Gulf of California

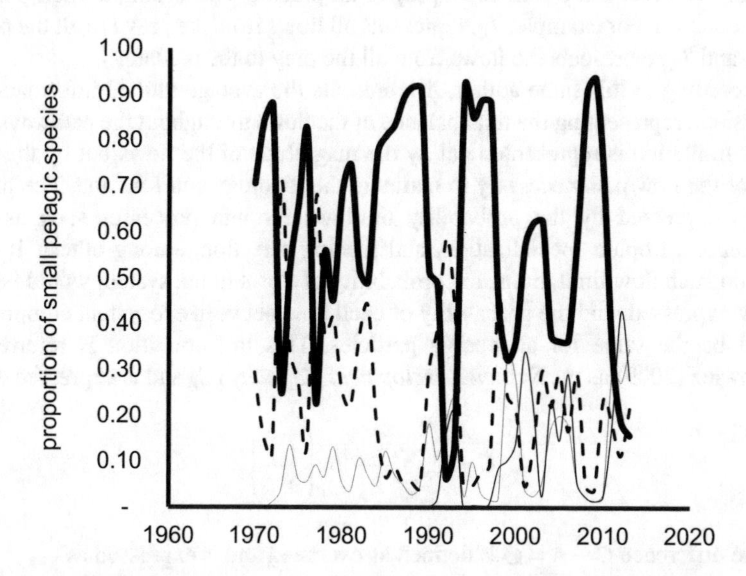

Fig. 2 Proportion of small pelagic species in commercial catches in the Gulf of California. Note the large fluctuations in the Monterey sardine (bold solid line) and the thread herring (bold dashed line) catches. The thin solid line represents other mixed species (chub mackerel, round herring, anchovy, and shortjaw leatherjack)

marine mammals, seabirds, elasmobranchs, polychaetes, zooplankton, meiobenthos, phytoplankton, shrimp trawl discards and detritus. The input data to the model, such as biomass, B, production/biomass P/B, consumption biomass Q/B, catches, diets, and discards, were used to obtain and parameterize a mass balance model and to verify its biological consistency (Annex). In the model, giant squids, shrimp trawling including discards, sardines, and the small-scale fleets were explicitly represented.

Entropy

According to Ulanowicz (1986, 2009), the organization of an ecosystem is defined by the *Ascendancy* (A) through the probabilities of energy flow between its components, which is represented as

$$A = T_{\circ,\circ} \sum_{i,j} \frac{T_{i,j}}{T_{\circ,\circ}} \log\left(\frac{T_{i,j} T_{\circ,\circ}}{T_{i,\circ} T_{\circ,j}}\right) \quad (1)$$

where $T_{i,j}$ represents the flow of a particle from the prey i to the predator j; the symbol $\circ$ represents the sum of all prey or all predators according to their position in the notation. For example, $T_{i,\circ}$ represents all flows from the prey i to all the predators j, and $T_{\circ,j}$ represents the flows from all the prey to the predator j.

According to this same author, A represents the average mutual information of the system, representing the final balance of the flows throughout the pathways. The flow's limitation is represented not by the magnitude of the flows but by the presence of the flow path necessary to maintain the topology and function. The limitation is expressed by the probability of flows through processes such as prey preferences, trophic specialization, and feeding selection, among others. If there were no such flow limitations, the probability of flows in the system would be randomly expressed, and the probability of each flow between ecosystem components would be the same for all energy particles. This last condition is referred by Ulanowicz (2009) as the *System Development Capacity* (C) and is expressed as

$$C = T_{\circ,\circ} \sum_{i,j} \frac{T_{i,j}}{T_{\circ,\circ}} \log\left(\frac{T_{i,j}}{T_{\circ,\circ}}\right) \quad (2)$$

The difference $C - A = \varnothing$ is defined as overhead and is expressed as

$$\varnothing = T_{\circ,\circ} \sum_{i,j} \frac{T_{i,j}}{T_{\circ,\circ}} \log\left(\frac{T_{i,j}^2}{T_{i,\circ} T_{\circ,j}}\right) \quad (3)$$

where $\varnothing$ represents the energy of the system in reserve.

According to Ulanowicz (2009), the A/C ratio is a relative measure of the organization of the ecosystem; then, $\left[1-\left(A/C\right)\right] = \varepsilon$ will be a measure of the relative entropy and is expressed as

$$\varepsilon = 1 - \sum_{i,j} \left[\frac{\left(T_{\circ,\circ}\right)^2}{T_{i,\circ} T_{\circ,j}} \right] \tag{4}$$

According to the previous definitions, Arreguín-Sánchez et al. (2020) propose a holistic indicator that represents the change in the ecosystem's entropy due to the effect of biomass removal from one or more of its components. The basic principle of this indicator, called a noxycline, is that when biomass is removed from a component of the ecosystem, a change in entropy is generated. If this removal of biomass is carried out systematically and increases over time, the change in the ecosystem's entropy will be gradual and have a certain tendency. As long as the capacity of the system to return to its initial state is not affected, it would be expected that by suspending the extraction of biomass, the system would recover its organization; if that threshold is exceeded, the system will lose that capacity. This threshold level would be observed as the turning point in the trend of entropy change when a certain harvest rate (removal of biomass) is applied that corresponds to the limit of removed biomass. Removal above this limit would cause an irreversible change to the ecosystem.

Noxycline Estimation

To estimate the noxycline (Arreguín-Sánchez et al., 2020) we developed a simulation experiment based on the trophic model of the Gulf of California ecosystem (Annex). Assuming a stable system, we brought the model to an initial state without exploitation. To do this, for each exploited functional group, the catch and discards were added to the corresponding live biomass, and fishing mortality was eliminated, leaving the P/B ratio represented only by natural mortality (Allen, 1971).

To simulate the change in entropy associated with gradually increasing biomass extraction, a period of 50 years was considered, increasing the harvest rate 2% each year such that, at the end of the period, only a remaining biomass of 2% remains. To obtain this effect, the harvest rate for each year, $\mathrm{HR}_{i,y}$, (Gulland, 1983) was represented as:

$$\mathrm{HR}_{i,y} = \frac{F_{i,y}}{M_{i,y} + F_{i,y}} \left[1 - e^{-\left(M_{i,y} + F_{i,y}\right)} \right] \tag{5}$$

where F_i is the instantaneous rate of fishing mortality on species i and year y, and represents the proportion of biomass removed by fishing with respect to the available biomass, or $\mathrm{HR}_{i,y} = C_{i,y}/B_{i,y}$; and $M_{i,y}$ represents the instantaneous rate of natural mortality for species i and year y, equivalent to the $\left(P/B\right)_{i,y}$ ratio in the absence of fishing.

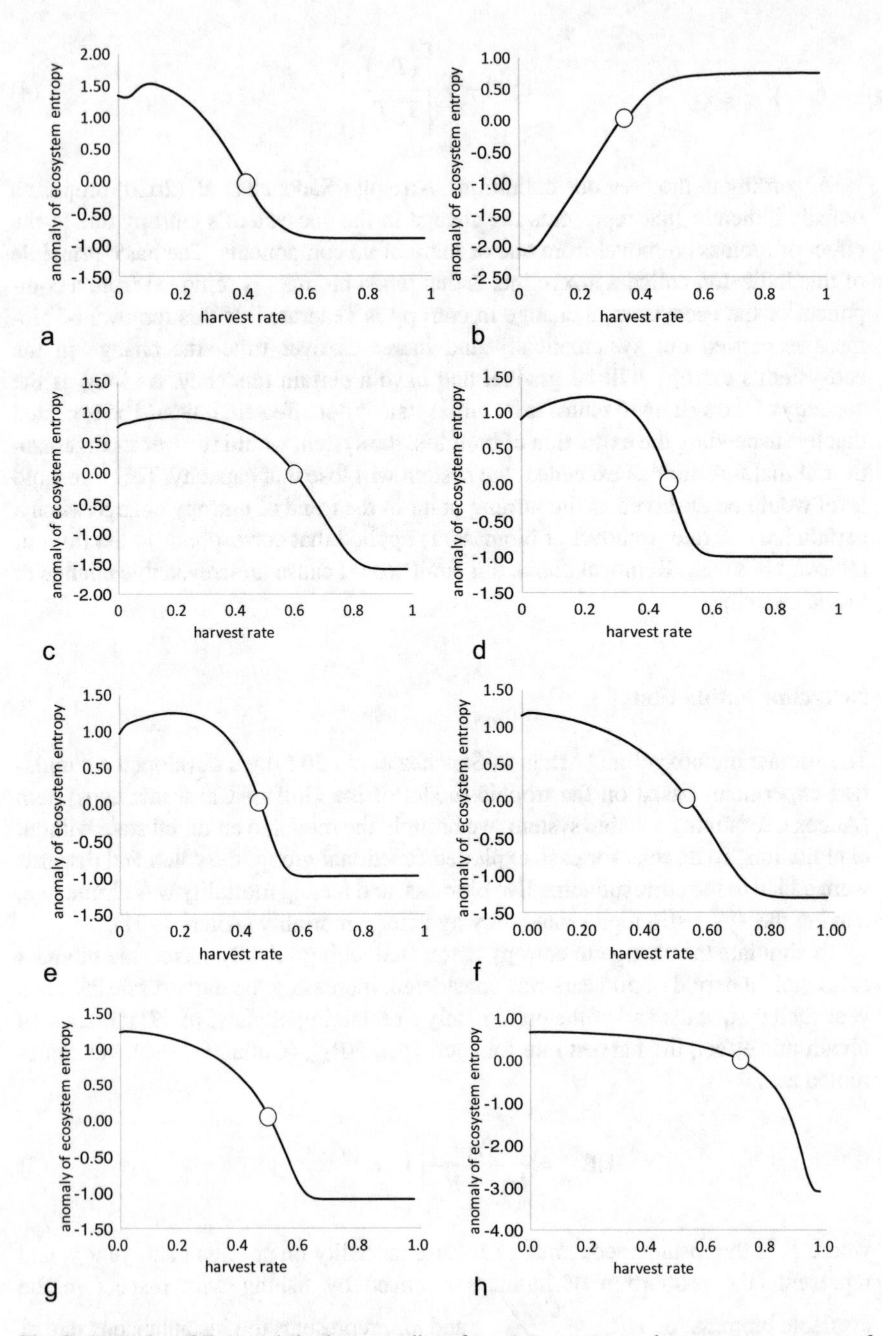

Fig. 3 Examples of the trends in the anomalies of ecosystem entropy changes across a range of gradually increasing harvest rates for several species in the central Gulf of California. The circle denotes the inflection point corresponding to the $ELRP_i$. (**a**) Marlin, (**b**) elasmobranch, (**c**) pineapple sardine, (**d**) giant squid, (**e**) Monterey sardine, (**f**) thread herring, (**g**) shrimp, (**h**) shortjaw leatherjack

Since the desired values of $HR_{i,y}$ and $M_{i,y}$ are known for each year from Eq. (5), it is possible to estimate $F_{i,y}$. Then, the pattern of $F_{i,y}$ is applied to the corresponding functional group i using the *Ecosim* model (Walters et al., 1997). After biomass was removed for some groups in the model at an increasing harvest rate year after year, the basic statistics on changes in *Ascendancy* and *System Development Capacity* were collected, and the gradual change in ecosystem entropy with the harvest rates was estimated. From the trend of entropy versus harvest rates, we observed the inflection point in the trajectory, identifying the values of $F_{i,y}$ and the harvest rate $HR_{i,y}$ corresponding to an ecosystem limit reference point associated with species i ($ELRP_i$). This procedure was run for each component group of the ecosystem trophic model, and the noxycline (Arreguín-Sánchez et al., 2020) was identified by linking the different $ELRP_i$ s of the different species.

Figure 3 shows examples for some clupeoids and other functional groups of the trajectories of the change in ecosystem entropy due to the extraction of biomass from each species by applying a harvest rate that gradually increases, showing the inflection point or the ecosystem limit reference point associated with each species i ($ELRP_i$). In Fig. 4, the noxycline is shown as a result of the tendency of all ELRPs.

The biological reference points, *BRPs*, used for the management of fisheries (Gabriel & Mace, 1999; Collie & Gislason, 2001) conventionally refer to populations, defining both target, BRP_{tar}, and limit, BRP_{lim}, reference points. The latter is particularly useful because it allows the identification of unwanted or high-risk conditions, facilitating the practice of precautionary management. According to the

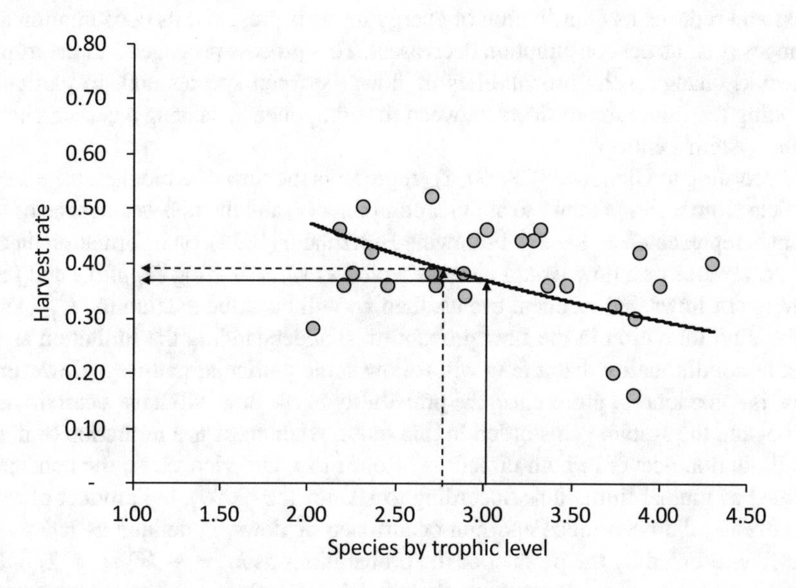

Fig. 4 Noxycline (black solid line) showing the trend of the inflection points of the change in ecosystem entropy caused by biomass removal for the functional groups of the Gulf of California. The arrows relate the noxycline to the species by their trophic level and harvest rate for the Monterey sardine and the thread herring, approximately showing the corresponding $ELRP_i$

characteristics of the exploited stock, *BRPs* can refer to various population attributes, such as the harvest rate required to obtain a catch volume equivalent to a given desired yield, a certain survival rate, a certain marginal yield, or a proportion of surviving spawning biomass, among others.

When trying to incorporate ecosystem-based management criteria in fishery studies using a population approach, information is available on the direct and indirect effects of fishing on the natural system, either regarding non-target species or habitat, and management measures are taken to reduce the undesirable effects (Worm & Lenihan, 2013). In the best case, the management measures will include the area of the natural system that is under the influence of the fishing activity and its species; however, management measures do not fully consider the effects on the function and organization of the ecosystem, as they are designed from the perspective of the population towards the ecosystem.

The noxycline, on the other hand, offers a holistic indicator that is defined through an attribute that involves the entire ecosystem. From the theoretical point of view, the greatest possible entropy of an ecosystem is when the probability that the particles of energy pass from one species to another is equal for all species and at any point in the ecosystem; that is, the components *i* and *j* are totally independent of each other. In contrast, the organization of the ecosystem reflects the conditionality or degree of limitation of flows between components. In terms of the flow paths, these limitations are translated as food preferences or some degree of specialization for prey. In this way, when biomass is removed from a species by fishing, the probabilities of flow between species are altered; the target species or stock loses biomass and reduces its contribution of energy towards prey, and its consumption also changes (i.e., its net consumption decreases). This process propagates in the trophic network, changing the probabilities of flows between species and, in particular, affecting the limitation of flows between the components, causing a certain change in the system's entropy.

According to Ulanowicz (2009), $T_{i,j}$ represents the flow of a biomass (or energy) particle from a component *i* to another component *j*, and the non-occurrence of that event is represented as $1 - T_{i,j}$. Following Boltzmann (1872), on information theory, the occurrence of a flow would be represented as $s_{i,j} = -k \log (T_{i,j})$. If *i* and *j* are, in terms of flows, independent events, then $s_{i,j}$ will be at the maximum, $s_{i,j}^{\max}$, since there is no limitation in the flow probabilities, understanding the limitation as the lack of conditionality that a flow will follow some particular pathway. However, if there is some kind of preference, the probability of the flow will have a certain limitation, and the system gains order. In this sense, when there is a limitation of flows, the limitation occurs in both directions, from *i* to *j*, and vice versa; the concept is defined as mutual limitation. According to Ulanowicz (2009), the product of these two events, the occurrence and non-occurrence of flows, is defined as indeterminacy, represented by the product of the probabilities as $h_{i,j} = -kT_{i,j} \log (T_{i,j})$. For the components of an ecosystem, the total indeterminacy, which expresses the capacity for the development of the system, is represented as the sum for all the components.

$$H = -k \sum_{i,j} T_{i,j} \log\left(T_{i,j}\right) \tag{6}$$

where if k is replaced by the total flows of the ecosystem $T_{\circ,\circ}$ to give a measure of the size of the system, and if $T_{i,j}$ is replaced by $T_{i,j}/T_{\circ,\circ}$, which represents the probability of flows, then Eq. (2) is obtained, which represents the development capacity.

According to the aforementioned definitions, the difference between $s_{i,j}^{\max} - s_{i,j} = x_{i,j}$ represents the mutual limitation. For the whole system, the average mutual limitation is represented as $X = \sum_{i,j} T_{i,j} x_{i,j}$; in terms of flow probabilities, for the entire ecosystem, the resulting expression is Eq. (1), which represents the Ascendancy (see Ulanowicz, 1986; Ulanowicz et al., 2009). Under the same considerations, when the events H and X are independent, then $X = H = 1$ or $A = C = 1$, and represents the entropy of the system, such that $C \geq A \geq 0$. Since the Ascendancy, A, represents the order of the system and the Development Capacity, C, represents the total indeterminacy, then A/C represents, in terms of proportion, a relative index of the ecosystem order. The difference in the unit will represent the opposite, the entropy, $\mathcal{E}$, as represented in Eq. (4). In this way, reducing the biomass of a component of the ecosystem reduces the limitation of flows represented by $s_{i,j}$; that is, the contribution of the removed biomass to the limitation of flows, and therefore to the order of the system, is reduced. This process does not occur through $s_{i,j}^{\max}$ since flows are represented by random probabilities, and a reduction in flows due to a reduction in biomass does not affect the randomness of the probabilities.

According to the above, at the inflection point, if the biomass reduction reaches the average mutual limitation to the limit of the ecosystem's balanced flows, then the probabilities of flows will change. The ecosystem will adopt a new state of average mutual limitation, and the previous order will be lost.

The noxycline represents a reference level of the ecosystem that is linked to its organization, which is reduced according to the amount of biomass extracted. The ELRP_i represents the critical limit of the ecosystem's balance flows that is required to maintain the ecosystem's order. In this context, the ELRP_i has a holistic nature that allows the determination of a harvest rate limit and represents a tool for management strategy. In effect, it is an approach that starts from the ecosystem and moves toward the individual resource.

In terms of the Gulf of California sardine case study, with the Monterrey sardine being the most abundant small pelagic species, the noxycline criterion allows us to estimate how much sardine biomass must remain in the sea after fishing to maintain the functioning and order of the ecosystem.

In this context, two particular attributes of the noxycline are worth mentioning: (1) the harvest rate (Eq. (5)) corresponding to the ELRP_i represents the proportion of the capture in terms of the available biomass; that is, $\text{HR}_i = C_i/B_i$. In this way, ELRP_i in terms of HR_i would be less sensitive to biomass changes; for example, if $\text{HR}_i = 0.3$, it would mean that the catch equals 30% of the biomass; if the biomass increases or decreases, HR_i would represent 30% of a higher biomass or lower biomass; and (2) the noxycline is only a limit reference point and does not correspond to a target harvest rate; consequently, it should essentially be considered as a precautionary criterion for management.

In our particular case study, for the two most abundant species, the Monterey sardine and the thread herring of the Gulf of California (which constitute 98% of the total catches in recent decades), the harvest rate corresponding to the $ELRP_i$ was near $HR_i = 0.38$ (Fig. 4). This means that, in practice, the biomass and availability of the fish stocks should be estimated every year before the beginning of the fishing season to establish the required biomass to remain in the sea after fishing. In this case, the remaining biomass would correspond to a minimum of 62% of the available biomass at the beginning of the season. This criterion will help to maintain fishing in a way that is compatible with the dynamic sustainability of the ecosystem.

Acknowledgements The authors thank Conservation and Biodiversity A.C., COBI, particularly María José Espinosa Romero, who promoted and supported this approach towards the sardine fishery of the Gulf of California. We thank the National Polytechnic Institute for support through the EDI and COFAA programs. Thanks to Dr. Diego Lercari-Bernier, Luis O. Duarte, Danilo Calliari Cuadro and Rodolfo Edward Vögler Santos for their valuable contributions in discussions on this issue. We also thank SEP-CONACyT (221705, A1-S-19598) and SIP (20180929, 20195986, 2019RE/124) for partial support.

Annex

Input data for the *Ecopath* model of the central Gulf of California

	Group name	B (t/km^2)	P/B (/year)	Q/B (/year)
1	Sea mammals	0.015	2.165	29.304
2	Sea birds	0.013	0.391	89.032
3	Sciaenidae	0.3087754	0.791	4.417
4	Scombridae	0.2952854	0.864	7.265
5	Sharks/Rays	0.1785297	0.465	9.48
6	Dolphin fish	0.36	0.833	3.4
7	Marlin	0.14	0.6	8
8	Sailfish	0.06	0.49	5
9	other billfish	0.08	0.68	7
10	Squid	1.139855	3.163	19.265
11	Carangidae	0.9207015	0.635	3.204
12	Serranidae	0.327	1.323	6.444
13	Scorp/Triglidae	0.112	0.821	3.883
14	Other fish	2.205302	1.528	7.954
15	Haemulidae	0.337	1.474	8.567
16	Monterrey sardine	3.666544472	3.309	10.556

	Group name	B (t/km²)	P/B (/year)	Q/B (/year)
17	Thread herring	0.021526548	1.48	11.873
18	Chub mackerel	0.038279712	1.885	4.947
19	Round herring	0.005596898	5.89	26.579
20	Anchovy	0.00502284	5.89	26.579
21	Pacific anchoveta	0.008058786	5.89	26.579
22	Shortjaw leatherjack	0.005300051	1.885	4.947
23	Lutjanidae	0.174161	0.601	5.088
24	Paralichthydae	0.5433912	0.584	2.987
25	Other molluscs	0.686	2.018	8.752
26	Myctophidae	0.8648831	1.456	7.937
27	Other macrocrus	0.4927337	2.337	7.938
28	Red crab	0.059	6.734	24.547
29	Shrimp	0.5020179	5.875	20.14
30	Crabs	0.3532665	2.638	9.072
31	Polychaeta	0.6201127	4.283	20.611
32	Stomatopods	0.01871599	2.168	8.747
33	Zooplankton	20.175	24.458	97.704
34	Meiobenthos	1.648046	8.444	52.928
35	Phytoplankton	26.14013	66.396	0
36	Discarded fish	0.647		
37	Detritus	2		

Commercial catches and discards (t/km²)

	Commercial catches per fleet				Discards
Group name	Small Scale	Shrimp Trawl	Sardina	Squid	Shrimp Trawl
Sciaenidae	0.0320				0.0252
Scombridae	0.0920				0.0006
Sharks/Rays	0.0610				0.0001
Squid				0.3670	
Carangidae	0.0010				0.0009
Serranidae	0.1800				0.1796
Scorp/Triglidae					0.0621
Other fish	0.0540				0.054
Haemulidae	0.0010				0.1955
Monterrey sardine			2.3510		0.0054
Thread herring			1.3803		
Chub mackerel			1.2272		
Round herring			0.1794		
Anchovy			0.1610		
Pacific anchoveta			0.9043		
Shortjaw leatherjack			0.1699		
Lutjanidae	0.0080				0.0001

	Commercial catches per fleet				Discards
Group name	Small Scale	Shrimp Trawl	Sardina	Squid	Shrimp Trawl
Paralichthydae	0.1260				0.1256
Myctophidae					0.0001
Shrimp		1.4320			
Crabs	0.2030				

Diet matrix for the ecosystem of the central Gulf of California

	Prey \ predator	1	2	3	4	5	6	7	8	9	10	11
1	Sea mammals											
2	Sea birds											
3	Sciaenidae					0.102						
4	Scombridae					0.090	0.128	0.342	0.010	0.304		
5	Sharks/Rays					–		0.004				
6	Dolphin fish						0.017	0.093		0.041		
7	Marlin							0.005	0.000			
8	Sailfish							0.011	0.005			
9	other billfish							0.007				
10	Squid	0.133	0.125			0.024	0.103	0.300	0.098	0.266	0.059	
11	Carangidae	0.118	0.101			0.004						
12	Serranidae											
13	Scorp/Triglidae				0.001	0.001						0.001
14	Other fish	0.167	0.167	0.042		0.028	0.047	0.112	0.241			
15	Haemulidae											
16	Monterrey sardine	0.345	0.367	0.403	0.523	0.331	0.174	0.118	0.190	0.224	0.508	0.370
17	Thread herring						0.001	0.000	0.001	0.001		
18	Chub mackerel						0.001	0.000	0.001	0.001		
19	Round herring						0.000	0.000	0.000	0.000		
20	Anchovy						0.001	0.000	0.001	0.001		
21	Pacific anchoveta						0.000	0.000	0.001	0.001		
22	Shortjaw leatherjack						0.000	0.000	0.000	0.000		
23	Lutjanidae					0.002						
24	Paralichthydae											
25	Other molluscs			0.041		0.128						0.038
26	Myctophidae			0.045	0.170	0.087	0.092		0.404			
27	Other macrocrus			0.091	0.004	0.010						
28	Red crab					0.062						
29	Shrimp			0.088		0.015					0.016	
30	Crabs					0.033						
31	Polychaeta											
32	Stomatopods					0.008						
33	Zooplankton	0.236	0.126	0.117	0.257	0.023	0.126		0.048	0.162	0.417	0.589
34	Meiobenthos											

	Prey \ predator	1	2	3	4	5	6	7	8	9	10	11
35	Phytoplankton			0.014								0.001
36	Discarded fish		0.113		0.031	0.052						0.001
37	Detritus			0.160	0.014							

	Prey \ predator	12	13	14	15	16	17	18	19	20	21	22
1	Sea mammals											
2	Sea birds											
3	Sciaenidae											
4	Scombridae											
5	Sharks/Rays											
6	Dolphin fish											
7	Marlin											
8	Sailfish											
9	other billfish											
10	Squid			0.061				0.020				
11	Carangidae	0.007		0.021								
12	Serranidae				0.010							
13	Scorp/Triglidae	0.001										
14	Other fish	0.183	0.129	0.061	0.342			0.310				
15	Haemulidae	0.007			0.012							
16	Monterrey sardine			0.207				0.170				
17	Thread herring											0.150
18	Chub mackerel											
19	Round herring											
20	Anchovy							0.230				0.150
21	Pacific anchoveta											
22	Shortjaw leatherjack											
23	Lutjanidae	0.012			0.021							
24	Paralichthydae	0.001		0.003								
25	Other molluscs	0.122										
26	Myctophidae											
27	Other macrocrus	0.104	0.106						0.100	0.100		
28	Red crab											
29	Shrimp	0.061	0.146	0.016	0.042			0.080				
30	Crabs	0.167	0.045									
31	Polychaeta	0.072		0.061				0.030				
32	Stomatopods	–	0.057									
33	Zooplankton	0.075		0.269	0.294	0.857	0.600	0.070	0.600	0.900	0.021	0.700
34	Meiobenthos	0.080	0.517		0.002							
35	Phytoplankton	0.074		0.145	0.161	0.143	0.400		0.300		0.979	
36	Discarded fish	0.025		0.010								
37	Detritus	0.007		0.147	0.115			0.090				

	Prey\predator	23	24	25	26	27	28	29	30	31	32	33	34
1	Sea mammals												
2	Sea birds												
3	Sciaenidae												
4	Scombridae												
5	Sharks/Rays												
6	Dolphin fish												
7	Marlin												
8	Sailfish												
9	other billfish												
10	Squid												
11	Carangidae	0.006											
12	Serranidae	0.002											
13	Scorp/Triglidae	0.002											
14	Other fish	0.042	0.080										
15	Haemulidae	0.001											
16	Monterrey sardine												
17	Thread herring												
18	Chub mackerel												
19	Round herring												
20	Anchovy												
21	Pacific anchoveta												
22	Shortjaw leatherjack												
23	Lutjanidae	0.002											
24	Paralichthydae												
25	Other molluscs					0.155							
26	Myctophidae			0.073									
27	Other macrocrus			0.089	0.021				0.003				
28	Red crab												
29	Shrimp	0.084	0.052						0.010				
30	Crabs	0.195							0.030				
31	Polychaeta		0.363			0.112	0.203						
32	Stomatopods												
33	Zooplankton	0.162		0.452	0.591		0.037					0.159	
34	Meiobenthos	0.209		0.022		0.293	0.153	0.310	0.293	0.249	0.200		0.042
35	Phytoplankton	0.187		0.365	0.358	0.155	0.005	0.178				0.779	
36	Discarded fish					0.001			0.001		0.062		
37	Detritus	0.109	0.505		0.030	0.284	0.602	0.512	0.663	0.751	0.738	0.062	0.958

Biological/ecological model consistency. Slopes of log (B), log (P/B), log (P/Q), log (P/R), log (R/B), and log (R/A) must be negative respect trophic levels (*B* biomass, *P* production, *Q* consumption, *R* respiration, *A* assimilation)

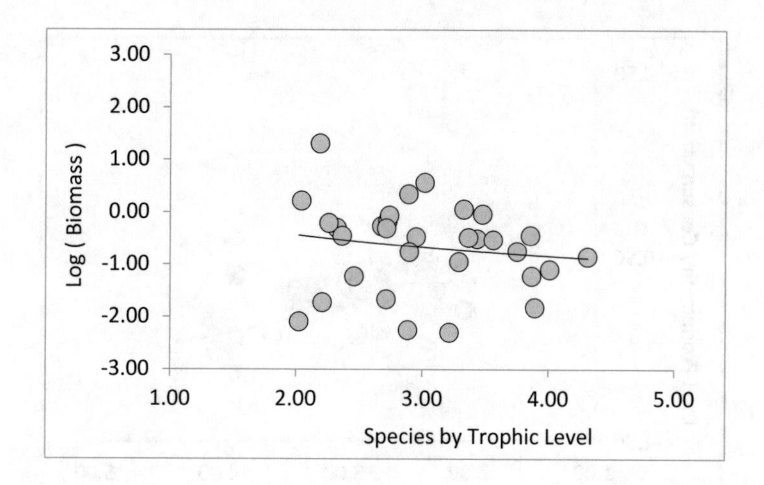

3.00
2.00
1.00
0.00
-1.00
-2.00
-3.00
Log (Biomass)
1.00
2.00
3.00
4.00
5.00
Species by Trophic Level

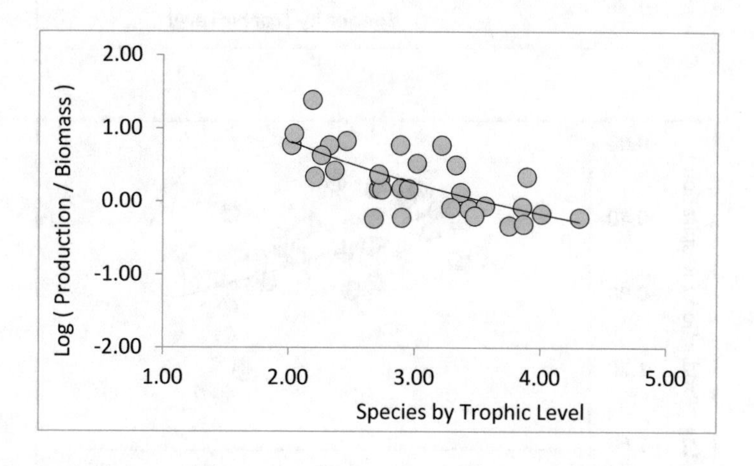

2.00
1.00
0.00
-1.00
-2.00
Log (Production / Biomass)
1.00
2.00
3.00
4.00
5.00
Species by Trophic Level

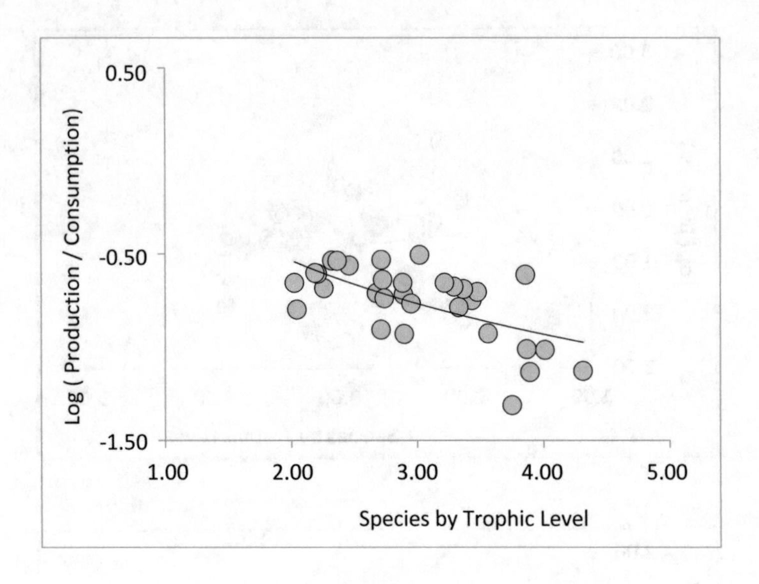
0.50
-0.50
-1.50
1.00
2.00
3.00
4.00
5.00
Log (Production / Consumption)
Species by Trophic Level

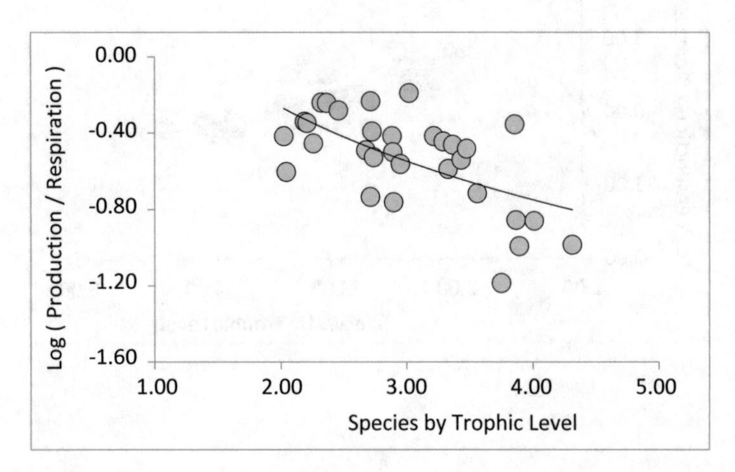
0.00
-0.40
-0.80
-1.20
-1.60
1.00
2.00
3.00
4.00
5.00
Log (Production / Respiration)
Species by Trophic Level

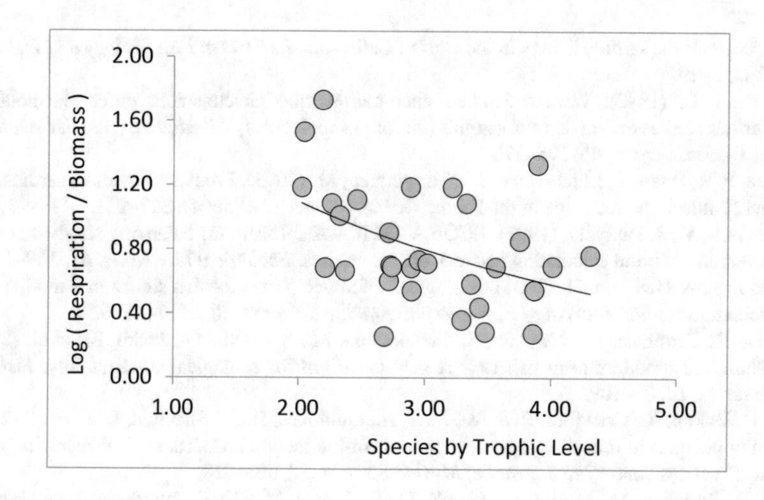

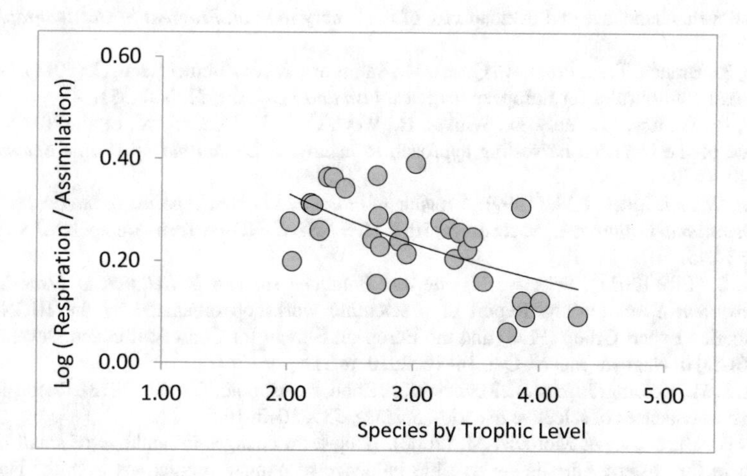

References

Allen, K. R. (1971). Relation between production and biomass. *Journal of the Fisheries Research Board of Canada, 28*, 1573–1581.

Arreguín-Sánchez, F., Arcos, E., & Chávez, E. A. (2002). Flows of biomass and structure in an exploited benthic ecosystem in the Gulf of California, Mexico. *Ecological Modelling, 156*, 167–183.

Arreguín-Sánchez, F., Del Monte-Luna, P., Zetina-Rejón, M.J., Duarte, L.O., Lercari, D., Riofrío-Lazo, M., et al. (2020). *Noxicline: reference limit for fish harvest based on ecosystem entropy*. Submitted to Fish and Fisheries.

Bakun, A., Babcock, E. A., Lluch-Cota, S. E., Santora, C., & Salvadeo, C. J. (2009). Issues of ecosystem-based management of forage fisheries in "open" non-stationary ecosystems: the example of the sardine fishery in the Gulf of California. *Reviews in Fish Biology and Fisheries, 20*(1), 9–29.

Boltzmann, L. (1872). Weitere Studien über das Wärme gleichgenicht unfer Gasmoläkuler [Further studies on the thermal equilibrium of gas molecules]. *Sitzungsberichte der Akademie der Wissenschaften, 66*, 275–370.

Chavez, F. P., Ryan, J., Lluch-Cota, S. E., & Ñiquen, M. (2003). From anchovies to sardines and back: Multidecadal change in the Pacific Ocean. *Science, 299*(5604), 217–221.

Christensen, V., & Pauly, D. (1992). ECOP A TH II -- A software for balancing steady-state ecosystem models and calculating network characteristics. *Ecological Modelling, 61*, 169–185.

Collie, J. S., & Gislason, H. (2001). Biological reference points for fish stocks in a multispecies context. *Canadian Journal of Fisheries and Aquatic Sciences, 58*, 2167–2176.

Colloca, F., Cardinale, M., Maynou, F., Giannoulaki, M., Scarcella, G., Jenko, K., et al. (2013). Rebuilding Mediterranean fisheries: A new paradigm for ecological sustainability. *Fish and Fisheries, 14*, 89–109.

Cury, P., Bakun, A., Crawford, R. J. M., Jarre, A., Quiñones, R. A., Shannon, L. J., et al. (2000). Small pelagics in upwelling systems: patterns of interaction and structural changes in "wasp-waist" ecosystems. *ICES Journal of Marine Science, 57*, 603–618.

Fréon, P., Bouchon, M., Mullon, C., García, C., & Ñiquen, M. (2008). Interdecadal variability of anchoveta abundance and overcapacity of the fishery in Peru. *Progress in Oceanography, 79*, 401–412.

Froese, R., Branch, T. A., Proel, A., Quaas, M., Sainsbury, K., & Zimmermann, C. (2011). Generic harvest control rules for European fisheries. *Fish and Fisheries, 12*, 340–351.

Froese, R., Walters, C., Pauly, D., Winker, H., Weyl, O. L. F., Demirel, N., et al. (2016). A critique of the balanced harvesting approach to fishing. *ICES Journal of Marine Science, 73*, 1640–1650.

Gabriel, W.L. & Mace, P.M. (1999). *A review of biological reference points in the context of the precautionary approach*. Proceedings, 5th NMFS NSAW. NOAA Tech. Memo. NMFS-F/SPO-40: 34:45.

García, S. (Ed). (2011). *Selective Fishing and Balanced Harvest in Relation to Fisheries and Ecosystem Sustainability*. Report of a scientific workshop organized by the IUCN-CEM Fisheries Expert Group (FEG) and the European Bureau for Conservation and Development (EBCD) in Nagoya (Japan), Oct. 14–16, 2010. p. 31.

Garcia, S. M., Kolding, J., Rice, J., Rochet, M. J., Zhou, S., Arimoto, T., et al. (2012). Reconsidering the consequences of selective fisheries. *Science, 335*, 1045–1047.

Grbec, B., Dulcic, J., & Morovic, M. (2002). Long-term changes in landings of small pelagic fish in the eastern Adriatic — possible influence of climate oscillations over the Northern Hemisphere. *Climate Research, 20*, 241–252.

Griffiths, S. P., Olson, R. J., & Watters, G. M. (2013). Complex wasp-waist regulation of pelagic ecosystems in the Pacific Ocean. *Reviews in Fish Biology and Fisheries, 23*, 459–475.

Gulland, J. A. (1983). *Fish stock assessment: A manual of basic methods*. New York: Wiley.

Hilborn, R., Amoroso, R. O., Bogazzia, E., Jensen, O. P., Parma, A. M., Szuwalski, C., et al. (2017). When does fishing forage species affect their predators? *Fisheries Research, 191*, 211–221.

Laugen, A. T., Engelhard, G. H., Whitlock, R., Arlinghaus, R., Dankel, D. J., Dunlop, E. S., et al. (2014). Evolutionary impact assessment: accounting for evolutionary consequences of fishing in an ecosystem approach to fisheries management. *Fish and Fisheries, 15*, 65–96.

Mansfield, B. (2011). "Modern" industrial fisheries and the crisis of overfishing. In R. Peet, P. Robbins, & M. Watts (Eds.), *Global political ecology* (pp. 84–99). London and New York: Routledge, Taylor & Francis Group.

Murawski, S. A. (2010). Rebuilding depleted fish stocks: The good, the bad, and, mostly, the ugly. *ICES Journal of Marine Science, 67*, 1830–1840.

Palkovacs, E. P. (2011). The overfishing debate: an eco-evolutionary perspective. *Trends in Ecology and Evolution, 26*(12), 616–617.

Pikitch, E. K. (2012). The risk of overfishing. *Science, 338*, 474–475.

Rosa, R., Gonzalez, L., Broitman, B. R., Garrido, S., Santos, A. M. P., & Nunes, M. L. (2010). Bioenergetics of small pelagic fishes in upwelling systems: relationship between fish condition, coastal ecosystem dynamics and fisheries. *Marine Ecology Progress Series, 410*, 205–218.

Roux, J. P., van der Lingen, C. D., Gibbons, M. J., Moroff, N. E., Shannon, L. J., Smith, A. D. M., et al. (2013). Jellyfication of marine ecosystems as a likely consequence of overfishing small pelagic fishes: Lessons from the Benguela. *Bulletin of Marine Science, 89*(1), 249–284.

Smith, A. D. M., Brown, C. J., Bulman, C. M., Fulton, E. A., Johnson, P., Kaplan, I. C., et al. (2011). Impacts of fishing low–trophic level species on marine ecosystems. *Science, 333*, 1147–1150.

Tommasi, D., Stock, C. A., Pegion, K., Vecchi, G. A., Methot, R. D., Alexander, M. A., et al. (2017). Improved management of small pelagic fisheries through seasonal climate prediction. *Ecological Applications, 27*(2), 378–388.

Tourre, Y. M., Lluch-Cota, S. E., & White, W. B. (2007). Global multi-decadal ocean climate and small-pelagic fish population. *Environmental Research Letters, 2*, 1–9.

Ulanowicz, R. E. (1986). *Growth and development: Ecosystem phenomenology*. New York: Springer-Verlag.

Ulanowicz, R. E. (2009). The dual nature of ecosystem dynamics. *Ecological Modelling, 220*, 1886–1892.

Ulanowicz, R. E., Goerner, S. J., Lietaer, B., & Gomez, R. (2009). Quantifying sustainability: Resilience, efficiency and the return of information theory. *Ecological Complexity, 6*, 27–36.

Walters, C., Christensen, V., & Pauly, D. (1997). Structuring dynamic models of exploited ecosystems from trophic mass-balance assessments. *Reviews in Fish Biology and Fisheries, 7*, 139–172.

Worm, B., & Branch, T. A. (2012). The future of fish. *Trends in Ecology and Evolution, 27*(11), 594–599.

Worm, B., & Lenihan, H. S. (2013). Threats to marine ecosystems: overfishing and habitat degradation. In M. D. Bertness, J. F. Bruno, B. R. Silliman, & J. J. Stachowicz (Eds.), *Marine community ecology and conservation* (pp. 449–476). Sunderland: Sinauer Associates.

Macroscopic Network Properties and Spatially-Explicit Dynamic Model of the Banco Chinchorro Biosphere Reserve Coral Reef (Caribbean Sea) for the Assessment of Harvest Scenarios

Fabián Alejandro Rodríguez-Zaragoza and Marco Ortiz

1 Introduction

Banco Chinchorro coral reef was declared a Biosphere Reserve in 1996 by the Mexican government to protect its biodiversity and ecosystem processes and to manage its natural resources (INE, 2000). This ecological system is located in the northern sector of the Mesoamerican Barrier Reef System and, at 40.7 km long and 18 km wide, is considered to be one of the largest platform coral reefs in the Caribbean Sea (Acosta-González, Rodríguez-Zaragoza, Hernández-Landa, & Arias-González, 2013; Jordán & Martín, 1987). The reef has high biodiversity due to its notable habitat heterogeneity, integrated into surrounding coral reefs are developments of spurs-and-groove habitats, wide stretches of seagrass and algae beds, coral reef patches, and small areas of mangrove (Acosta-González et al., 2013).

However, the reef has historically been exploited by artisanal fishers (>40 years), whose main target species are spiny lobster (*Panulirus argus*), queen conch snail (*Lobatus gigas*), and several fish species (Sosa-Cordero, 2003). As a consequence of this intensive period of harvest, *L. gigas* and *P. argus* are currently considered to

F. A. Rodríguez-Zaragoza (✉)
Laboratorio de Ecología Molecular, Microbiología y Taxonomía (LEMITAX), Departamento de Ecología, Centro Universitario de Ciencias Biológicas y Agropecuarias (CUCBA), Universidad de Guadalajara, Zapopan, Jalisco, Mexico
e-mail: fabian.rzaragoza@academicos.udg.mx

M. Ortiz
Departamento de Biología Marina, Facultad de Ciencias del Mar, Universidad Católica del Norte, Coquimbo, Chile

Laboratorio de Modelamiento de Sistemas Ecológicos Complejos (LAMSEC), Instituto Antofagasta (IA), Universidad de Antofagasta, Antofagasta, Chile

Instituto de Ciencias Naturales Alexander von Humboldt, Facultad de Ciencias Naturales y Recursos Biológicos, Universidad de Antofagasta, Antofagasta, Chile

M. Ortiz, F. Jordán (eds.), *Marine Coastal Ecosystems Modelling and Conservation*, https://doi.org/10.1007/978-3-030-58211-1_8

be over-exploited resources (Cala de la Hera, de Jesús-Navarrete, Oliva-Rivera, & Ocaña-Borrego, 2012; de Jesús-Navarrete, Medina-Quej, & Oliva-Rivera, 2003; De Jesús-Navarrete & Valencia-Hernández, 2013; Sosa-Cordero, 2003). In order that these stocks may recover, the Mexican government has established minimum extraction sizes and bans for the fishing cooperatives on the exploitation of *L. gigas* and *P. argus* (de Jesús-Navarrete et al., 2003; Rodríguez-Zaragoza et al., 2016). However, poaching activity has caused the situation to reach a critical state, negatively affecting the livelihoods of legal fishers (de Jesús-Navarrete et al., 2003). While the bans have been implemented, the exploitation of reef fish of the Serranidae, Lutjanidae, and Haemulidae families on Banco Chinchorro has increased considerably, impacting ecosystem functioning and properties (resistance) (Rodríguez-Zaragoza et al., 2016). An additional perturbation is the introduction and rapid spread of the alien lionfish *Pterois volitans* (Ortiz et al., 2015) since its presence may decrease the overall biodiversity of coral reefs and lead to phase-shift transitions from corals to fleshy macroalgae (Albins & Hixon, 2013).

Studies regarding fishing activities have mainly focused on the exploitation of *P. argus* and *L. gigas* using classical population analysis (de Jesús-Navarrete et al., 2003; Sosa-Cordero, 2003). Some spatially-explicit predictions and habitat classification models have shown that fish diversity hotspots are highly correlated with reefscapes composed of an aggregation of coral colonies with seagrass beds (Acosta-González et al., 2013). Besides, qualitative and quantitative ecosystem models have been built for analyzing management strategies in Banco Chinchorro from an ecosystem perspective. Rodríguez-Zaragoza et al., (2016) built several stationary trophic models to analyze the multispecies fishery, the structure, trophic functioning, and ecosystem growth and development of five subsystems at Banco Chinchorro reef. Their outcomes showed that, as a consequence of the ecological heterogeneity of this coral reef, a subsystem-level management strategy needs to be designed, particularly because different species or functional groups exhibit a greater sensitivity to human interventions depending on which area they inhabit.

Nowadays, the Ecosystem-Based Fisheries Management (EBFM) is a widely recognized and accepted analytical strategy to assess multispecies fisheries (Pikitch et al., 2004), incorporating the needs of the authorities, fishers, tourism service operators, and others involved, and ensuring the implementation of a holistically sustainable co-management strategy (Ortiz et al., 2013, 2015). Ecosystem mass-balance models may be considered as complementary tools for studies of population dynamics. These models can be constructed using the program *EcopathWithEcosim* (EwE) (Christensen & Walters, 2004), integrating fishing activities, diet matrices, and network analysis. EwE incorporating the *Ecospace* routine has frequently been used to build spatially-explicit models based on multitrophic relationships, assessing the possible effects of applying different management strategies in marine ecosystems (i.e. Walters, Christensen, & Pauly, 1997; Walter, Pauly, Christensen, & Kitchell, 1999; Ortiz and Wolff, 2002; Ortiz, Avendaño, Berrios & Campos 2009; Ortiz, Avendaño, Cantillañez, Berrios & Campos, 2010; Romagnonia, Mackinsonb, Hong & Eikeset 2015; Alexander, Meyjes & Heymans 2016). Nevertheless, few *Ecospace* models have been built specifically for coral reefs (Gribble, 2005; Okey et al., 2004; Varkey, Ainsworth, & Pitcher, 2012). Recognizing that *Ecospace* models enable the propagation of higher-order effects as a response to fishing activities to be assessed across spatial scales within marine ecosystems, the main objective of this chapter was to build a mass-balance model using EwE that incorporates the spatial heterogeneity of the coral reef at Banco Chinchorro Biosphere Reserve. To achieve this the five subsystems or habitat types previously described were considered (Rodríguez-Zaragoza et al., 2016), permitting us to assess: (1) biomass distribution and determination of the macroscopic properties of the whole ecological system; (2) spatial changes as responses to the eventual application of different fishing scenarios on commercially interesting species; and (3) the species or functional groups that are most impacted by different spatially-explicit management scenarios.

Banco Chinchorro is a coral reef with an ovoid shape (43.2 km long x 18.0 km wide) and platform type, located off the south-west coast of Yucatán Peninsula and separated from the continent by a channel 30.8 km wide and ≈500 m deep (INE, 2000; Vega-Zepeda, Hernández-Arana, & Carricart-Ganivet, 2007) (Fig. 1a). This coral reef has a lagoon with an area >500 km^2 and depths varying between 1 and 9 m, surrounded by a semi-continuous barrier reef (~115 km in perimeter), where the seawater is oligotrophic with average surface water temperatures that range between 27 and 29 °C, while salinity varies from 36.6‰ to 36.9‰ (INE, 2000). More details regarding the environmental features of this coral reef are described in Ortiz et al. (2015) and Rodríguez-Zaragoza et al., (2016).

2 Modeling Strategy and Assumptions

EcopathWithEcosim (EwE) was initially based on the Polovina (1984) approach, which estimates the biomass and food consumption of several functional groups within an ecosystem. Subsequently, Christensen and Pauly (1992) and Walters et al. (1997) made some extensions to EwE, increasing its capabilities to allow simulations of temporal (i.e. *Ecosim*) and spatial (i.e. *Ecospace*) dynamics. EwE permits steady-state ecosystem models to be assessed in terms of matter/energy flow at a particular time, whereas the *Ecospace* routine provides temporal dynamic simulations of *Ecopath*, where biomass (B) and consumption (Q) dynamics are evaluated in spatial and temporal dimensions, this means that they vary within the spatial coordinates *x*, *y*, and over time (Fig. 1b). For more details of *Ecospace* theoretical framework see Box 1. Moreover, EwE also includes a network analysis feature called *Ascendency* (Ulanowicz, 1986, 1997), which allows us to estimate macroscopic properties, such as growth, organization, development, and the "ecosystem health." In this context, an ecosystem would be considered healthy if it is sustainable because it keeps its organization and processes over time, and is resilient against disturbances (Costanza, Mageau, & Norton, 1998). For more details about *Ascendency* see chapter "Macroscopic Properties and Keystone Species Complexes in Kelp Forest Ecosystems Along the North-Central Chilean Coast."

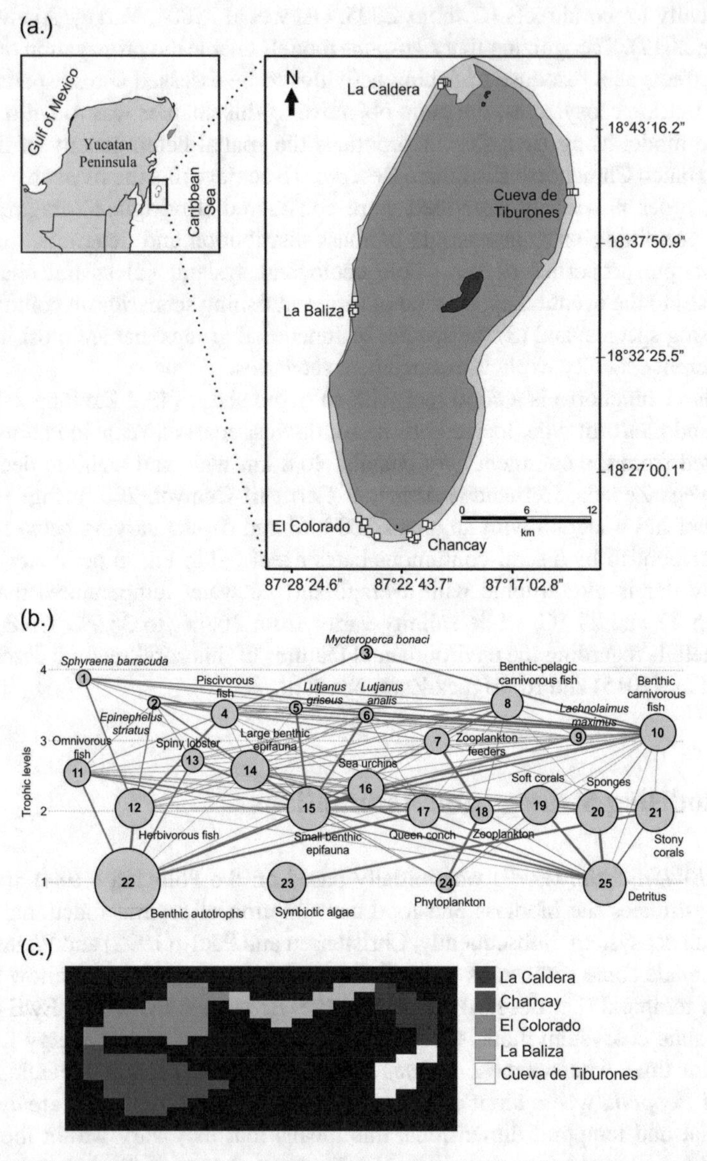

Fig. 1 (**a**) Study area at Banco Chinchorro Biosphere Reserve, Mexico. (**b**) Trophic model for Banco Chinchorro. Vertical position approximates trophic level. The circle size is proportional to the compartment (species and functional groups) biomass (g wet weight [ww] m^{-2}). The connections represent the flow of matter among compartments. The number in circle corresponds to the species or functional groups (for details see Table 1). (**c**) Spatial basemap constructed through *Ecospace* routine of EwE, showing the five different subsystems, and (**d**) Spatial fishing effort scenarios simulated by *Ecospace*. (In the subsystem Cueva Tiburones: fishing on Spiny lobster; in La Caldera: fishing on *S. barracuda*, *E. striatus*, *M. bonaci*, *L. analis*, BPCF, *L. maximus*, BCF; in La Baliza: fishing on *S. barracuda*, *E. striatus*, *M. bonaci*, *L. griseus*, *L. analis*, BPCF, *L. maximus*, BCF, Spiny lobster, Queen conch; in El Colorado: fishing on *S. barracuda*, *E. striatus*, *M. bonaci*, PF, *L. griseus*, *L. analis*, *L. maximus*, BCF, Spiny lobster; in El Chankay: fishing on *E. striatus*, *M. bonaci*, *L. maximus*)

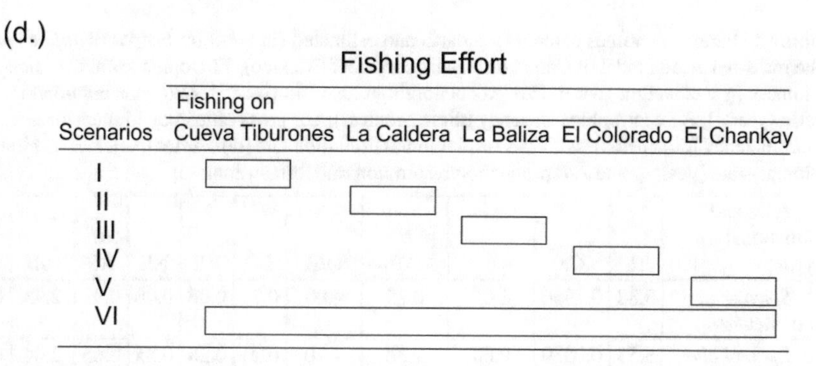

Fig. 1 (continued)

Box 1 Ecospace Theoretical Framework

The *Ecospace* is a spatially-explicit routine of *EcopathWithEcosim* program that permits us to define rectangular grids of spatial cells. In this case, the space, time, and state of variables are considered discrete by using the Eulerian approach that considers movement as flow of organisms among fixed cells. The immigration rate by cell is assumed to consist of four emigration flows from the surroundings cells. The emigration flows are represented as instantaneous movement rates (m_i) *x* biomass (B_i) in each cell as follows:

$$B_i = m_{i(x,y)} * B_{i(x,y)} \tag{1}$$

where (x,y) represents cell row and column.

Likewise, fishing mortality (F_i) can be spatially represented by using a gravity function incorporated into *Ecospace*, by which the proportion of total effort allocated to each cell is considered to be proportional to the sum over groups of biomass multiplied by catchability and market price of the commercial species or functional groups, all is integrated by following algorithm:

$$G_{kc} = \frac{O_{kc} * U_{kc}\left(\sum_i^n p_{ki} * q_{ki} * B_{ic}\right)}{C_{kc}} \tag{2}$$

where G_{kc} is weighted attractiveness of cell *c* to fleet *k*; $O_{kc} = 1$ if cell *c* is open to the fleet and 0 if it is closed to fishing; $U_{kc} = 1$ if it was specified that gear *k* can harvest and 0 otherwise; p_{ki} is the relative price assigned for species or functional group *i* by fleet *k* fisheries; q_{ki} is the catchability of compartment *i* by fleet *k*; B_{ic} is the biomass of species or group *i* in cell *c*; and C_{kc}

Table 1 Parameter values entered (standard) and estimated (in bold) by *EcopathWithEcosim* for the mass-balanced model of Chinchorro Bank coral reef (Mexico). *TL* trophic level, *Ca* catches, *B* biomass [g wet weight (ww)], *P/B* production/biomass ratio [year^{-1}], *Q/B* consumption/biomass ratio [year^{-1}], *EE* ecotrophic efficiency [dimensionless], *GE* gross efficiency [dimensionless], *NE* net efficiency [dimensionless], *R/AS* respiration/assimilation rate [dimensionless], *R/B* respiration/biomass rate [year^{-1}], and *P/R* production/respiration rate [dimensionless]

Species and functional groups	TL	Ca	B	P/B	Q/B	EE	GE	NE	R/AS	R/B	P/R
1. *Sphyraena barracuda*	3.88	0.1950	1.05	0.25	4.00	0.74	0.06	0.08	0.92	2.95	0.08
2. *Epinephelus striatus*	3.53	0.0530	0.63	1.32	4.70	0.24	0.28	0.35	0.65	2.44	0.54
3. *Mycteroperca bonaci*	4.25	0.0430	0.70	0.37	3.40	0.17	0.11	0.14	0.86	2.35	0.16
4. Piscivorous fish	3.38	0.0002	7.91	1.16	13.20	0.25	0.09	0.11	0.89	9.40	0.12
5. *Lutjanus griseus*	3.47	0.0140	0.70	0.54	9.10	0.96	0.06	0.07	0.93	6.74	0.08
6. *Lutjanus analis*	3.37	0.4640	0.88	0.58	5.20	0.91	0.11	0.14	0.86	3.58	0.16
7. Zooplankton feeders (ZF)	3.00		5.80	2.50	14.06	0.92	0.18	0.22	0.78	8.75	0.29
8. Benthic-pelagic carnivorous fish (BPCF)	3.53	0.0050	19.20	0.26	8.39	0.98	0.03	0.04	0.96	6.45	0.04
9. *Lachnolaimus maximus*	3.06	0.0580	1.21	0.56	4.78	0.97	0.12	0.15	0.85	3.26	0.17
10. Benthic carnivorous fish (BCF)	3.12	0.0450	37.60	2.30	9.98	0.99	0.23	0.29	0.71	5.68	0.40
11. Omnivorous fish (OF)	2.55		8.54	1.88	38.35	0.93	0.05	0.06	0.95	36.47	0.05
12. Herbivorous fish (HF)	2.04		57.70	1.49	24.49	0.97	0.06	0.07	0.94	23.00	0.06
13. Spiny Lobster	2.73	0.7340	4.10	1.10	7.40	0.97	0.15	0.15	0.85	6.30	0.17
14. Large benthic epifauna (LBE)	2.59		50.20	2.10	7.50	0.94	0.28	0.35	0.65	3.90	0.54
15. Small benthic epifauna (SBE)	2.05		114.80	6.95	40.85	0.88	0.17	0.18	0.83	33.90	0.21
16. Sea urchins	2.32		36.00	1.10	3.80	0.93	0.29	0.36	0.64	1.94	0.57
17. Queen conch	2.00	0.0890	18.90	1.64	14.00	0.25	0.12	0.15	0.85	9.56	0.17

(continued)

Table 1 (continued)

Species and functional groups	TL	Ca	B	P/B	Q/B	EE	GE	NE	R/AS	R/B	P/R
18. Zooplankton	2.00		4.45	40.00	165.00	0.95	0.24	0.30	0.70	92.00	0.43
19. Soft corals	2.09		50.80	1.09	9.00	0.96	0.12	0.14	0.88	7.91	0.14
20. Sponges	2.00		102.50	1.40	5.20	0.93	0.27	0.34	0.66	2.76	0.51
21. Stony corals	2.00		54.40	1.09	9.00	0.97	0.12	0.13	0.88	7.91	0.14
22. Benthic autotrophs (BA)	1.00		4992.98	13.25		0.04					
23. Symbiotic algae	1.00		54.11	10.20		0.66					
24. Phytoplankton	1.00		2.10	1185.00		0.51					
25. Detritus	1.00		118.00			0.07					

is the relative cost of fishing in cell *c* by gear *k*. Finally, the spatial simulation searches for a moving equilibrium for the biomass of each compartment based on the following function:

$$B_{i(t+\Delta t)} = W_{i(t)} * B_{i(t)} + \left(1 - W_{i(t)}\right) * B_{i(e)} \tag{3}$$

where $B_{i(t+\Delta t)}$ is the biomass of the compartment *i* moving toward an equilibrium along the time; $B_{i(t)}$ is the biomass of the compartment *i* at the initial time of simulation; $B_{i(e)}$ is the biomass of the compartment *i* at equilibrium; and $W_{i(t)}$ is the exponential weight for the compartment *i* and assumes the following behavior:

$$W_{i(t)} = e^{-(Z_i + E_i) * \Delta t} \tag{4}$$

where Z_i is the total instantaneous mortality rate for the compartment *i* and E_i is the total instantaneous emigration rate. For more details on Ecospace framework see Walter et al. (1999).

2.1 Data Sources, Model Compartments, and Dynamic Simulations

A global trophic model was constructed of the whole Banco Chinchorro coral reef with compartments representing species and functional groups following the criteria established by Rodríguez-Zaragoza et al. (2016). The functional fish groups were characterized as benthic-pelagic carnivorous fishes, piscivorous fishes, benthic carnivorous fishes, zooplankton feeders, omnivorous fishes, and herbivorous fishes. Other functional groups were the large benthic epifauna, sea urchins, soft corals, small benthic epifauna, zooplankton, stony corals, sponges, benthic autotrophs, symbiotic algae (*Symbiodinium* spp.), phytoplankton, and detritus. The species were selected for their economic importance: the queen conch snail *L. gigas*, the spiny lobster *P. argus,* and the reef fish *Mycteroperca bonaci*, *Sphyraena barracuda*, *Epinephelus striatus*, *Lutjanus griseus*, *Lutjanus analis*, and *Lachnolaimus maximus*. (For more details of the species, functional groups, and sampling procedures in the current study, please see Rodríguez-Zaragoza et al. (2016).

During the balancing process, the model was checked based on the following six guidelines proposed by Heymans et al. (2016): (1) The Ecotrophic Efficiency (EE) of all compartments had to be <1.0 (Ricker, 1968), (2) the Gross Efficiency (GE) of all compartments had to be <0.3 (Christensen & Pauly, 1993). If any inconsistencies were detected, the average biomass was modified within the confidence limits (±1 standard deviation), (3) the Net Efficiency of all compartments had to be >GE, (4) the Respiration/Assimilation (R/AS) had to be <1.0, (5) the Respiration/Biomass (R/B) values for fishes had to be 1–10 year^{-1} or, for groups with higher turnover, 50–100 year^{-1}, and (6) Production/Respiration (P/R) had to be <1.0 (Table 1).

The *Ecospace* simulations were performed using EwE software *v*. 6.4.1. Dispersal rates ranged from 300 km year^{-1} for fishes to 1.0 km year^{-1} for species that lived in only one subsystem and for sessile organisms, set based on personal observations made during fieldwork and from the range given by Ortiz et al. (2010) and Varkey et al. (2012). The relative dispersal values in poor habitats (i.e. unsuitable for the taxa) were the highest (factor = 10) for mobile consumers, such as most species and functional groups of fish, medium (factor = 5–8) for spiny lobster, large benthic epifauna, small benthic epifauna, zooplankton, and phytoplankton, and lowest (factor = 2–4) for slow and sessile species or functional groups. Relative vulnerability to predation in poor habitats ranged from 2.0 for top predators (e.g. *S. barracuda* and *M. bonaci*) to 100.0 for the species and functional groups of lower trophic levels. Relative feeding rate in poor habitats ranged from 1.0 for top predators, plankton and detritus, to 0.01–0.02 for slow motion and sessile organisms. For all other components an intermediate value (0.5) was used. The spatial distribution of each subsystem in the study area is shown in Fig. 1c. Several fishery scenarios were evaluated over a five-year period, where only the impact on the four most important species was considered in terms of catch and demand (spiny lobster, queen conch, *S. barracuda,* and *L. analis*). Spatially-explicit simulations were performed considering exclusive harvest from each subsystem, as well as simultaneous harvests in all subsystems (Fig. 1d). *Ecospace* simulations were conducted based on three flow controls (i.e. different vulnerabilities, (v_{ij})) that affect the energy transfer rate between two compartments. The following flow controls were used: bottom-up (v = 1.0), mixed (v = 3.0), and top-down (v = 5.0). This approach was used because of the lack of the fishery data's time-series, making it is impossible to calibrate the *EwW* model. Market prices and operational costs were not included in the spatial simulations.

3 Macroscopic Network Properties and Dynamic-Spatial Model Responses

The functional groups of benthic autotrophs (BA), small benthic epifauna (SBE), and sponges comprised of the highest biomass of the entire Banco Chinchorro reef (Table 1). The high biomass magnitude for BA has been reported previously for other Mexican coral reefs (Acosta-González et al., 2013; Arias-González, González-Gándara, Cabrera, & Christensen, 2011; Arias-González, Nuñez-Lara, González-Salas, & Galzin, 2004). The size of the autotroph biomass has been conjectured to be a consequence of the lower herbivore pressure exerted by sea urchins, the impact of fishing on large herbivores and the increase in sediments and nutrients from run-off in the seawater (Hughes, 1994; Jackson et al., 2001; Hughes et al., 2003; Fung, Seymour & Johnson, 2011; Arias-González et al., 2017). Similarly, the BA accounted for the highest values of *Total System Throughput* (TST) and *Ascendency* (A). However, the fish *M. bonaci* presented the lowest percentage of *Average Mutual Information* (AMI), which accounts for the complexity in the entire system (Table 2).

Regarding ecosystem growth and development, the size of the TST for Banco Chinchorro reef was higher than those reported for other coral reef systems, such as those in the Indo-Pacific (Arias-González, Delesalle, Salvat, & Galzin, 1997; Arias-

Table 2 Network flow indices for the ecological system of Banco Chinchorro coral reef (Mexico) after steady-state mass trophic model by *EcopathWithEcosim*. The units are given in g wet weight (g ww) and Flowbit is the product of flow (g ww m^{-2} $year^{-1}$) and bits

Network flow indices	
Total system throughput (TST) (g ww m^{-2} $year^{-1}$)	144,980.70
Ascendency (A) (g ww m^{-2} $year^{-1}$* bits)	184,988.00
Overhead (Ov) (g ww m^{-2} $year^{-1}$*bits)	119,299.50
Development capacity (C) (g ww m^{-2} $year^{-1}$*bits)	304,287.40
Average mutual information (AMI) (dimensionless)	1.28
M. bonaci is accounting for the lowest % of AMI	0.000037
Pathway redundancy (of internal flows of Overhead) (%)	44.54
A/C (%)	40.69
Ov/C (%)	59.31
Finn's cycling index (FCI) (%)	0.32
Finn's mean path length (FPL) (dimensionless)	2.09
Food web connectance (FWC) (dimensionless)	0.25
Omnivory Index (OI) (dimensionless)	0.11
Mean trophic level of the catch (dimensionless)	3.09

González & Morand, 2006; Liu et al., 2009), Eastern Tropical Pacific (Okey et al., 2004), and the Caribbean Sea (Arias-González et al., 2004; Opitz, 1996; Rodríguez-Zaragoza, 2007). However, our results were similar to those described for the Mahahual and Yuyum reefs located off the Mexican Caribbean coast (facing Banco Chinchorro) (Rodríguez-Zaragoza, 2007) (Table 2). The high biomass of the BA could explain the large size of TST. The *Ascendency*, *Overhead*, *Development Capacity*, *A/C*, and *Ov/C* ratios indicated that Banco Chinchorro reef would be a more developed, organized, and healthy ecological system compared to other coastal ecosystems (Baird and Ulanowicz, 1993; Wolff, 1994; Heymans and Baird, 2000; Wolff et al., 2000; Ortiz & Wolff, 2002; Arias-González et al., 2004, 2011; Arias-González & Morand, 2006; Ortiz, 2008; Cáceres et al., 2016; Ortiz et al., 2010, 2015, 2016), but also that this system was less resistant to perturbations. The latter factor could be explained by the lower harvest pressure exerted on this ecosystem.

With regard to the food web structure, Finn's cycling index (FCI), Finn's mean path length (FPL), and food web connectance (FWC) were calculated for Banco Chinchorro reef and were higher than those described for some Mexican Caribbean coastal reefs (Rodríguez-Zaragoza, 2007). Nevertheless, the system omnivory index (OI) for the system revealed similar magnitudes compared to models constructed for other coral reefs ecosystems (Arias-González & Morand, 2006), coastal lagoons (Vega-Cendejas & Arreguín-Sánchez, 2001), and benthic communities of temperate systems (Ortiz, 2008; Ortiz et al., 2010; Taylor, Wolff, Mendo, & Yamashiro, 2008). The impact of fishing on the network showed that the mean trophic level of catch in this study was similar to those described for other coral reefs (Arias-González et al., 2004; Liu et al., 2009; Rodríguez-Zaragoza, 2007) and mainly indicated exploitation of organisms from high and intermediate trophic levels. This outcome suggests that the fisheries of Banco Chinchorro reef have not yet generated severe disturbance to the ecosystem, such as would be the case should there be fishing down the food web process, which occurs when there is a considerable reduction in the population size of the large predatory fishes at the top of the food webs, as has been observed in other marine ecosystems (González, Torruco-Gomez, Liceaga-Correa, & Ordaz, 2003; Pauly, Christensen, Dalsgaard, Froese, & Torres, 1998).

The spatially dynamic simulations showed quite similar qualitative and quantitative patterns of direct and indirect effects on the remaining compartments using mixed and top-down flow control mechanisms. Conversely, the magnitude of changes using bottom-up flow control was markedly lower. According to the fishing model scenarios, the subsystems Cueva Tiburones, La Caldera, and El Chankay propagated the highest effects on the other components in the system, thus the harvest trajectory in these areas should be monitored. Likewise, fishing simultaneously in the five subsystems would not spread the greatest impact across the entire ecosystem; therefore, a harvest rotation policy would not be advisable (Fig. 2). It is relevant here to indicate that the validity of these findings is difficult to evaluate because only a few *Ecospace* models have been constructed for cross-checking between observed and predicted results. Despite this limitation, the dynamic model presented in the current study should be considered as a general (qualitative) strategy for examining the consequences of spatially-explicit fishing pressure, which could be useful for the design of sustainable multispecies fisheries management (Pauly et al., 2002), particularly considering that protected marine areas could restore the populations and ecological networks of adjacent highly exploited systems (Arias-González et al., 2004).

Although we are well aware that the quantitative trophic model constructed and analyzed in this study was a partial representation of the overall trophic makeup and interactions underlying the dynamics within the Banco Chinchorro reef ecosystem, such limitations are common in any type of model and independent of the model´s degree of complexity (Levins, 1966; Ortiz and Levins, 2011, 2017). In the current model, the following constraints were identified: (1) system complexity was reduced concerning the composition of several functional groups, although the most abundant species were considered; (2) regardless of the inherent well-known limitations and shortcomings of the *Ecopath* and *Ecosim* theoretical frameworks (Christensen & Walters, 2004), and recognizing that ecological processes occur in changing environments (Levins, 1968), the constructed model and its spatially-explicit simulations represented underlying system processes only when considering their short-term or transient dynamics (Ortiz, 2018; Ortiz et al., 2013, 2015, 2017).

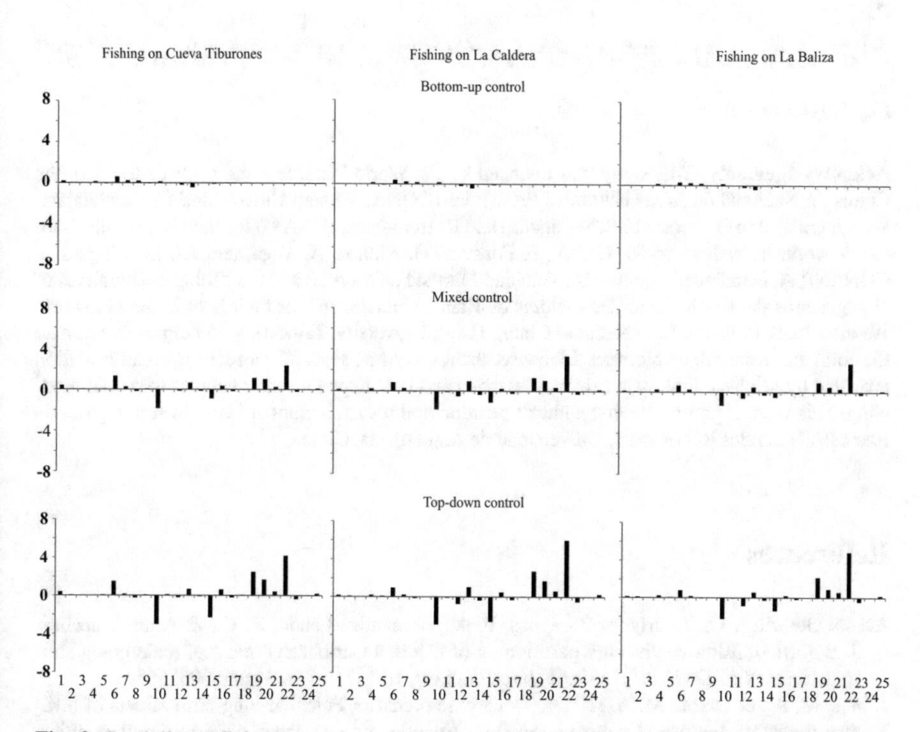

Fig. 2 Spatially-explicit propagation of direct and indirect effects after 5 years of simulation under six harvest scenarios using *Ecospace* routine of EwE in each subsystem. All simulations were done using bottom-up (v = 1.0), mixed (v = 3.0), and top-down (v = 5.0) flow control mechanism

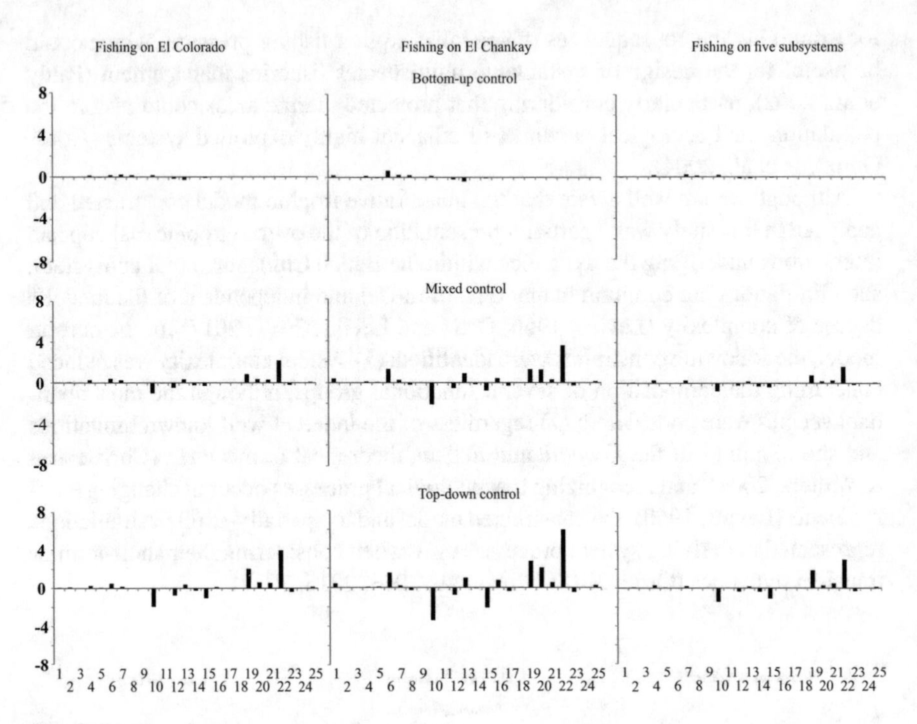

Fig. 2 (continued)

Acknowledgements This study was financed by the World Wildlife Fund (WWF)-México, the Comisión Nacional de Areas Naturales Protegidas (CONANP), and Universidad de Guadalajara. We are grateful to G. Acosta (CICY-Cancún) and R. Hernández (UNAM) for their help in the laboratory work; as well as to M. García, F. Fonseca, G. Muñoz, A. Vega, and J.J. Domínguez at CONANP, A. Hernándezs-Flores (Universidad Martista de Mérida) and the fishing cooperatives of "Langosteros del Caribe" and "Pescadores de Banco Chinchorro" for their help in the fieldwork. We also thank to Pi-Jen Liu (National Chung Hsing University, Taiwan), who help us to estimate the coral and macroalgae biomass. Likewise, thanks to A. de Jesús-Navarrete for the information provided by *L. gigas*. Finally, we thank F. Berrios and L. Campos for their support in spatial modeling. This work is part of the first author's postdoctoral research grant at Instituto Antofagasta de Recursos Naturales Renovables, Universidad de Antofagasta, Chile.

References

Acosta-González, G., Rodríguez-Zaragoza, F. A., Hernández-Landa, R. C., & Arias-González, J. E. (2013). Additive diversity partitioning of fish in a Caribbean coral reef undergoing shift transition. *PLoS ONE, 8*(6), e65665. https://doi.org/10.1371/journal.pone.0065665.

Albins, M. A., & Hixon, M. A. (2013). Worst case scenario: Potential long-term effects of invasive predatory lionfish (*Pterois volitans*) on Atlantic and Caribbean coral-reef communities. *Environmental Biology of Fishes, 96*, 1151–1157.

Alexander, K. A., Meyjes, S. & Heymans, J. J. (2016). Spatial ecosystem modelling of marine renewable energy installations: Guaging the utility of Ecospace. *Ecological Modelling, 331*, 115–128.

Arias-González, J. E., Delesalle, B., Salvat, B., & Galzin, R. (1997). Trophic functioning of the Tiahura sector Moorea Island, French Polynesia. *Coral Reefs, 16*, 231–246.

Arias-González, J. E., González-Gándara, C., Cabrera, J. L., & Christensen, V. (2011). Predicted impact of the invasive lionfish *Pterois volitans* on the food web of a Caribbean coral reef. *Environmental Research, 111*, 917–925.

Arias-González, J. E., & Morand, S. (2006). Trophic functioning with parasites: a new insight for ecosystem analysis. *Marine Ecology Progress Series, 320*, 43–53.

Arias-González, J. E., Nuñez-Lara, E., González-Salas, C., & Galzin, R. (2004). Trophic models for investigation of fishing effect on coral reef ecosystems. *Ecological Modelling, 172*, 197–212.

Arias-González, J. E., Fung, T., Seymour, R. M., Garza-Pérez, J. R., Acosta-González, G., Bozec, Y. M. & Johnson C. R. (2017). A coral-algal phase shift in Mesoamerica not driven by changes in herbivorous fish abundance. *PLOS ONE, 12*, e0174855.

Baird, D. & Ulanowicz, R. (1993). Comparative study on the trophic structure, cycling and ecosystem properties of four tidal estuaries. *Marine Ecology Progress Series, 99*, 221–237.

Cáceres, I., Ortiz, M., Cupul-Magaña, A. L. & Rodríguez-Zaragoza F. A. (2016). Trophic models and short-term simulations for the coral reefs of Cayos Cochinos and Media Luna (Honduras): a comparative network analysis, ecosystem development, resilience and fishery. *Hydrobiologia, 770*, 209–224.

Cala de la Hera, Y. R., de Jesús-Navarrete, A., Oliva-Rivera, J. J., & Ocaña-Borrego, F. A. (2012). Auto-ecology of the queen conch (*Strombus gigas* L. 1758) at Cabo Cruz, Eastern Cuba: Management and sustainable use implications. *Proceedings of the Gulf and Caribbean Fisheries Institute, 64*, 342–348.

Christensen, V., & Pauly, D. (1992). Ecopath II: A software for balancing steady-state ecosystem models and calculating network characteristics. *Ecological Modelling, 61*, 169–185.

Christensen, V., & Pauly, D. (1993). Trophic models of aquatic ecosystems. *ICLARM Conference Proceedings, 26*, 338–352.

Christensen, V., & Walters, C. (2004). Ecopath with Ecosim: Methods, capabilities and limitations. *Ecological Modelling, 172*, 109–139.

Costanza, R., Mageau, M., & Norton, B. (1998). Patten B predictors of ecosystem health. In D. Rapport, R. Costanza, P. Epstein, C. Gaudet, & R. Levins (Eds.), *Ecosystem health* (pp. 240–250). Malden: Blackwell Science.

de Jesús-Navarrete, A., Medina-Quej, A., & Oliva-Rivera, J. J. (2003). Changes in the queen conch (*Strombus gigas* L.) population structure at Banco Chinchorro,Quintana Roo, México, 1990–1997. *Bulletin of Marine Science, 73*, 219–229.

De Jesús-Navarrete, A., & Valencia-Hernández, A. (2013). Declining densities and reproductive activities of the queen conch *Strombus gig*as (Mesogastropoda: Strombidae) in Banco Chinchorro, Eastern Caribbean, Mexico. *Revista de Biología Tropical, 61*, 1671–1679.

Fung, T., Seymour, R. M. & Johnson, C. R. (2011). Alternative stable states and phase shifts in coral reefs under anthropogenic stress. *Ecology, 92*, 967–982.

González, A., Torruco-Gomez, D., Liceaga-Correa, A., & Ordaz, J. (2003). The shallow and deep bathymetry of the Banco Chinchorro reef in the Mexican Caribbean. *Bulletin of Marine Science, 73*, 15–22.

Gribble, N. A. (2005). *Ecosystem modelling of the great barrier reef: A balanced trophic biomass approach* (pp. 2561–2567). Melbourne: Modelling and Simulation Society of Australia and New Zealand.

Heymans, J. & Baird, D. (2000). A carbon flow model and network analysis of the northern Benguela upwelling system, Namibia. *Ecological Modelling, 126*, 9–32.

Heymans, J. J., Coll, M., Link, J. S., Mackinson, S., Steenbeek, J., Walters, C. & Christensen, V. (2016). Best practice in Ecopath with Ecosim food-web models for ecosystem-based management. *Ecological Modelling, 331*, 173–184.

Hughes, T. P. (1994). Catastrophes, phase shifts, and large-scale degradation of a Caribbean coral reef. *Science, 265*, 1547–1551.

Hughes, T. P., Baird, A. H., Bellwood, D. R., Card, M., Connolly, S. R., Folke, C., Grosberg, R., Hoegh-Gulberg, O., Jackson, J. B. C., Kleypas, J., Lough, J. M., Marshall, P., Myström, N., Palumbi, S. R., Pandolfi, J. M., Rosen, B., & Roughgarden, J. (2003). Climate change, human impacts, and the resilience of coral reefs. *Science, 301*, 929–933.

INE. (2000). *Programa de Manejo Reserva Banco Chinchorro*. México: Instituto Nacional de Ecología.

Jackson, J. B. C., Kirby, M. X., Berger, W. H., Bjordal, K. A., Botsford, L. W., Bourque, B. J., Bradbury, R. H., Cooke, R., Erlandson, J., Estes, J. A., Hughes, T. P., Kidwell, S., Lange, C. B., Lenihan, H. S., Pandolfi, J. M., Peterson, C. H., Steneck, R. S., Tegner, M. J., & Warner, R. R. (2001). Historical overfishing and the recent collapse of coastal ecosystems. *Science, 293*, 629–638.

Jordán, E., & Martín, E. (1987). Banco Chinchorro: Morphology and composition of a Caribbean atoll. *Atoll Research Bulletin, 310*, 1–25.

Levins, R. (1966). The strategy of model building in population biology. Princeton Monographs Series.

Levins, R. (1968). *Evolution in changing environments*. Princeton, NJ: Princeton University.

Liu, P. J., Kwang-Tsao, S., Rong-Quen, J., Tung-Yung, F., Saou-Lien, W., Jiang-Shiou, H., Jen-Ping, C., Chung-Chi, C., & Hsing-Juh, L. (2009). A trophic model of fringing coral reefs in Nanwan Bay, southern Taiwan suggests overfishing. *Marine Environmental Research, 68*, 106–117.

Okey, T. A., Banks, S., Born, A. F., Bustamante, R. H., Calvopiña, M., Graham, J. E., Espinoza, E., Fariña, J. M., Garske, L. E., Recke, G. K., Salazar, S., Shepherd, S., Toral-Granda, V., & Wallem, P. (2004). A trophic model of a Galápagos subtidal rocky reef for evaluating fisheries and conservation strategies. *Ecological Modelling, 172*, 383–401.

Opitz, S. (1996). *Trophic interactions in Caribbean coral reefs*. Manila, Philippines: ICLARM.

Ortiz, M. (2008). Mass balanced and dynamics simulations of trophic models of kelp ecosystems near the Mejillones Peninsula of northern Chile (SE Pacific): Comparative network structure and assessment of harvest strategies. *Ecological Modelling, 216*, 31–46.

Ortiz, O. (2018). Robustness of macroscopic-systemic network indices after disturbances on diet-community matrices. *Ecological Indicators, 95*, 509–517.

Ortiz., M. & Wolff M. (2002). Trophic models of four benthic communities in Tongoy Bay (Chile): comparative analysis and preliminary assessment of management strategies. *Journal of Experimental Marine Biology and Ecology, 268*, 205–235.

Ortiz M. & Levins, R. (2011). Re-stocking practices and illegal fishing in northern Chile (SE Pacific coast): a study case. *Oikos, 120*, 1402–1412.

Ortiz, M. & Levins, R. (2017). Self-feedbacks determine the sustainability of human interventions in eco-social complex systems: Impacts on biodiversity and ecosystem health. *PLoS ONE*; 12, e0176163.

Ortiz, M., Avendaño, M., Berrios, F. & Campos, L. (2009). Spatial and mass balanced trophic models of La Rinconada Marine Reserve (SE Pacific coast), a protected benthic ecosystem: management strategy assessment. *Ecological Modelling, 220*, 3413–3423.

Ortiz, M., Avendaño, M., Cantillañez, M., Berrios, F., & Campos, L. (2010). Trophic mass balanced models and dynamic simulations of benthic communities from La Rinconada Marine Reserve off northern Chile: Network properties and multispecies harvest scenario assessments. *Aquatic Conservation: Marine and Freshwater Ecosystems, 20*, 58–73.

Ortiz, M., Campos, L., Berrios, F., Rodríguez-Zaragoza, F. A., Hermosillo, B., & González, J. (2015). Mass balanced trophic models and short-term dynamical simulations for benthic ecological systems of Mejillones and Antofagasta bays (SE Pacific): Comparative network structure and assessment of human impacts. *Ecological Modelling, 309–310*, 153–152.

Ortiz, M., Levins, R., Campos, L., Berrios, F., Campos, F., Jordán, F., González, J., & Rodríguez-Zaragoza, F. A. (2013). Identifying keystone trophic groups in benthic ecosystems: implications for fisheries Management. *Ecological Indicators, 25*, 133–140.

Ortiz, M., Berrios F., González, J., Rodríguez-Zaragoza, F. & Gómez I. (2016). Macroscopic network properties and short-term dynamic simulations in coastal ecological systems at Fildes Bay (King George Island, Antarctica). *Ecological Complexity, 28*, 145–157.

Ortiz, M., Hermosillo-Nuñez, B., Gonzáleza, J., Rodríguez-Zaragoza, F., Gómez, I., Jordán, F. (2017). Quantifying keystone species complexes: Ecosystem-based conservation management in the King George Island (Antarctic Peninsula). *Ecological Indicators, 81*, 453–460.

Pauly, D., Christensen, V., Dalsgaard, J., Froese, R., & Torres, F. (1998). Fishing down marine food webs. *Science, 279*, 860–863.

Pauly, D., Christensen, V., Guénette, S., Pitcher, T. J., Sumaila, U. R., Walters, C. J., Watson, R., & Zeller, D. (2002). Towards sustainability in world fisheries. *Nature, 418*, 689–695.

Pikitch., E. K., Santora, C., Babcock, E. A., Bakun, A., Bonfil, R., Conover, D. O., Dayton, P., Doukakis, P., Fluharty, D., Heneman, B., Houde, E. D., Link, J., Livingston, P. A., Mangel, M., McAllister, M. K., Pope, J. & Sainsbury K. J. (2004). Ecosystem-Based Fishery Management. *Science, 305*, 346–347.

Polovina, J. J. (1984). Model of a coral reef ecosystem: I the Ecopath model and its application to French Frigate Shoals. *Coral Reefs, 3*, 1–11.

Ricker, W. E. (1968). Food from the sea. In Committee on Resources and Man (Ed.), *Resource and man* (pp. 87–108). San Francisco: US National Academy of Sciences/W. H. Freeman.

Rodríguez-Zaragoza, F. A. (2007). *Biodiversidad y funcionamiento de los ecosistemas arrecifales costeros del Caribe mexicano*. Ph.D. thesis, Centro de Investigación y Estudios Avanzados, Unidad-Mérida, México, p. 344.

Rodríguez-Zaragoza, F. A., Ortiz, M., Berrios, F., Campos, L., de Jesús-Navarrete, A., Castro-Pérez, J., Hernández-Flores, A., García-Rivas, M., Fonseca-Peralta, F. & Gallegos-Aguilar E. (2016). Trophic models and short-term dynamic simulations for benthic-pelagic communities at Banco Chinchorro Biosphere Reserve (Mexican Caribbean): a conservation case. *Community Ecology, 17*, 48–60.

Romagnonia, G., Mackinsonb, S., Hong, J. & Eikeset A. M. (2015). The Ecospace model applied to the North Sea: Evaluating spatial predictions with fish biomass and fishing effort data. *Ecological Modelling, 300*, 50–60.

Sosa-Cordero, E. (2003). Trends and dynamics of the spiny lobster, *Panulirus argus*, resource in Banco Chinchorro, Mexico. *Bulletin of Marine Science, 73*, 203–217.

Taylor, M. H., Wolff, M., Mendo, J., & Yamashiro, C. (2008). Changes in trophic flow structure of Independence Bay (Peru) over an ENSO cycle. *Progress in Oceanography, 79*, 336–351.

Ulanowicz, R. (1986). *Growth and development: Ecosystems phenomenology*. New York: Springer.

Ulanowicz, R. (1997). *Ecology, the ascendant perspective. Complexity in ecological systems series*. New York: Columbia University Press.

Varkey, D., Ainsworth, C. H., & Pitcher, T. J. (2012). Modelling reef fish population responses to fisheries restrictions in marine protected areas in the coral triangle. *Journal of Marine Biology, 2012*, 721483. https://doi.org/10.1155/2012/721483.

Vega-Cendejas, M. E., & Arreguín-Sánchez, F. (2001). Energy fluxes in a mangrove ecosystem from a coastal lagoon in Yucatan Peninsula, Mexico. *Ecological Modelling, 137*, 119–133.

Vega-Zepeda, A., Hernández-Arana, H., & Carricart-Ganivet, J. P. (2007). Spatial and size-frequency distribution of *Acropora* (Cnidaria: Scleractinia) species in Chinchorro Bank, Mexican Caribbean: implications for management. *Coral Reefs, 26*, 671–676.

Walter, C. J., Pauly, D., Christensen, V., & Kitchell, J. (1999). Representing density dependent consequences of the life history strategies in aquatic ecosystems: ECOSIM II. *Ecosystems, 3*, 70–83.

Walters, C. J., Christensen, V., & Pauly, D. (1997). Structuring dynamic models of exploited ecosystems from trophic mass-balance assessments. *Reviews in Fish Biology and Fisheries, 7*, 139–172.

Wolff, M. (1994). A trophic model for Tongoy Bay – a system exposed to suspended scallop culture (Northern Chile). *Journal of Experimental Marine Biology and Ecology, 182*, 149–168.

Wolff, M., Koch, V. & Isaac, V. (2000). A trophic flow model of the Caeté mangrove estuary (North Brazil) with considerations for the sustainable use of its resources. *Estuarine, Coastal and Shelf Science, 50*, 789–803.

The Use of Ecological Networks as Tools for Understanding and Conserving Marine Biodiversity

Viviana Márquez-Velásquez, Rafael L. G. Raimundo, Ricardo de Souza Rosa, and Andrés F. Navia

1 Introduction

The integrative understanding and management of anthropogenic and ecological processes that drive marine biodiversity are key to conserving marine ecosystems (Dunne et al., 2016; Estes, Heithaus, McCauley, et al., 2016; Lotze, Coll, & Dunne, 2011; Worm, Barbier, Beaumont, et al., 2006). A variety of anthropogenic pressures

V. Márquez-Velásquez
Fundación colombiana para la investigación y conservación de tiburones y rayas, Squalus, Colombia

Programa de Pós-Graduação em Ciências Biológicas (Zoologia), Universidade Federal da Paraíba, João Pessoa, PB, Brazil

Grupo de Pesquisa Ecologia Evolutiva de Interações Ecológicas: Redes, Teoriase Aplicações, Universidade Federal da Paraíba, Rio Tinto, PB, Brazil

R. L. G. Raimundo
Grupo de Pesquisa Ecologia Evolutiva de Interações Ecológicas: Redes, Teorias e Aplicações, Universidade Federal da Paraíba, Rio Tinto, PB, Brazil

Departamento de Engenharia e Meio Ambiente, Centro de Ciências Aplicadas e Educação, Universidade Federal da Paraíba, Rio Tinto, PB, Brazil

Programa de Pós-Graduação em Ecologia e Monitoramento Ambiental (PPGEMA), Universidade Federal da Paraíba, Rio Tinto, PB, Brazil

R. de Souza Rosa
Departamento de Sistemática e Ecologia, Centro de Ciências Exatas e da Natureza, Universidade Federal da Paraíba, João Pessoa, Brazil

A. F. Navia (✉)
Fundación colombiana para la investigación y conservación de tiburones y rayas, Squalus, Colombia

Grupo de investigación en Ecología Animal, Universidad del Valle, Cali, Colombia
e-mail: anavia@squalus.org

M. Ortiz, F. Jordán (eds.), *Marine Coastal Ecosystems Modelling and Conservation*, https://doi.org/10.1007/978-3-030-58211-1_9

threaten marine biodiversity by changing patterns of species richness, ecological interactions, and associated ecosystem functions at multiple spatial and temporal scales. Such pressures include overfishing (Jackson, Kirby, Berger, et al., 2001; McCauley et al., 2015; Pauly & Zeller, 2016), pollution (Islam & Tanaka, 2004; Nixon, 1995), introductions of alien species (Bax, Williamson, Aguero, Gonzalez, & Geeves, 2003; Vitousek, D'antonio, Loope, Rejmanek, & Westbrooks, 1997), and climate change (Hillebrand et al., 2018; Hoegh-Guldberg & Bruno, 2010). In addition, extinction of native species due to overfishing changes patterns of species interactions, affecting community stability and threatening the long-term persistence of marine biodiversity (Gilljam, Curtsdotter, & Ebenman, 2015; Jennings & Kaiser, 1998; Lotze et al., 2011).

The urgent need of mitigating anthropogenic impacts on the biological diversity of oceans led to novel methodological approaches aimed to bridge theoretical community ecology and conservation strategies aimed to protect marine biodiversity. Knowledge of ecological and evolutionary processes that shape food webs—which are networks defined by trophic interspecific interactions—is the cornerstone of system-based conservation approaches aimed to ensure the long-term persistence of functionally diverse marine ecological communities (Bascompte, Melián, & Sala, 2005; Dunne, Williams, & Martinez, 2002; Navia, Cortés, Jordán, Cruz-Escalona, & Mejía-Falla, 2012). On the way to gaining insight into food web organization and its consequences for community dynamics and ecosystem functioning, numerous controversies have emerged among researchers, not so much as to whether humans drive deleterious effects on food webs, but rather about their magnitude and the levels of biological organization affected by such effects (Navia, Cortés, & Cruz-Escalona, 2012). On the one hand, intense fishing pressures over the past 50 years have arguably reshaped the most fundamental properties of marine food webs, including species richness, abundances, trait-values distributions, and patterns of ecological interactions (Dunne et al., 2016; Jennings, Greenstreet, & Reynolds, 1999; Lotze et al., 2011; Pauly, Christensen, Dalsgaard, Froese, & Torres, 1998). On the other hand, ongoing anthropogenic changes may mostly affect species targeted by fisheries but do not necessarily imply community-level degradation (Essington, Beaudreau, & Wiedenmann, 2006; Litzow & Urban, 2009). Despite such controversy, growing evidence supports that anthropogenic activities rapidly reshape patterns of feeding interactions at the whole-community (network) level and lead to the rewiring of marine food web architecture just a few years after commercial exploitation starts (Baum & Worm, 2009; Ritchie & Johnson, 2009). As anthropogenic changes in marine communities affect species abundances and trait diversity, they threaten the stability, functioning, and persistence of marine ecosystems (Gilljam et al., 2015; Jennings & Kaiser, 1998; Lotze et al., 2011).

In the last decades, the structural analysis of ecological networks has emerged as a powerful tool for fisheries management and conservation planning (Dunne et al., 2002; Navia, Cruz-Escalona, Giraldo, & Barausse, 2016; Ortiz, Rodriguez-Zaragosa, Hermosillo-Nunez, & Jordán, 2015; Solé & Montoya, 2001; Stouffer & Bascompte, 2011). Empirical knowledge on food web structure and theoretical works on how topological properties relate to network persistence have unraveled fundamental relationships between biodiversity structure and ecosystem functioning (Estes et al., 2016; Lotze et al., 2011; Worm et al., 2006; Yen, Cabral, Cantor, et al., 2016). Network analyses also allow us to identify the topological role that each species plays within a food web, which can be used as a proxy for its contribution to community dynamics and hence inform conservation prioritization strategies (e.g., Bascompte et al., 2005; Dunne, Williams, & Martinez, 2004; Jordán, 2009; Jordán, Liu, & Mike, 2009; Luczkovich, Borgatti, Johnson, & Everett, 2003; Rezende, Albert, Fortuna, & Bascompte, 2009). Such advances in ecological network theory can greatly help us to understand, predict, and manage the effects of anthropogenic impacts on marine biodiversity (Dunne et al., 2016).

Beyond structural analyses, dynamic network modeling using computer simulations can provide theoretically founded predictions on how biodiversity loss is expected to reshape community resilience and functional diversity and hence influence the long-term persistence of ecosystems (Dunne et al., 2002, 2016; Raimundo, Guimarães Jr, & Evans, 2018). Network models can greatly benefit from species-interaction data widely available for different types of ecosystems (e.g., Bornatowski, Navia, & Barreto, 2017; Bornatowski, Navia, Braga, Abilhoa, & Corrêa, 2014; Endrédi, Jordán, & Abonyi, 2018; Gaichas & Francis, 2008; Marina et al., 2018; Navia, Cortés, & Mejía-Falla, 2010; Navia et al., 2016; Navia, Maciel-Zapata, González-Acosta, Leaf, & Cruz-Escalona, 2019). In addition to informing models of community dynamics, ecological network data can enhance strategies of environmental education, as they allow the visualization of ecological communities, a level of biological organization that is a key for biodiversity maintenance but rarely addressed by education professionals outside specialist circles. By incepting the community ecology perspective into conservation and environmental education programs, the network approach can also greatly contribute to public engagement through communication, citizen science, and evidence-based advocacy for decision-makers (Pocock et al., 2016).

In this chapter, we discuss network tools commonly used for understanding the architecture of marine biodiversity and for predicting how food webs are expected to respond to alternative management strategies, such as reducing fishing pressure on particular target species, setting fish catch quotas, or selectively removing particular species from the system.

2 Background

2.1 Advances in Network Ecology and Their Relevance for Conservation

The network approach to complex systems has been widely applied in ecology, providing synthetic metrics to describe patterns of interspecific interactions that characterize community structure (Montoya, Pimm, & Solé, 2006; Strogatz, 2001).

Species-interaction networks capture the fundamental relationship between the architecture of biodiversity and ecosystem functions, whose understanding can improve our ability to predict how anthropogenic perturbations change biodiversity dynamics (Dunne et al., 2002; Jordano, 2016). Empirical knowledge of ecological network structure (e.g., Bascompte & Jordano, 2014; Pascual & Dunne, 2006) associated with a diversity of structural analyses (e.g., Olesen, Bascompte, Dupont, & Jordano, 2007) and network modeling approaches (e.g., Allesina & Tang, 2015) is shedding light on mechanisms by which species interactions modulate ecological and evolutionary dynamics at the whole-community level (e.g., Guimarães Jr., Pires, Jordano, Bascompte, & Thompson, 2017; Thébault & Fontaine, 2010). The ecological network approach is feeding integration between ecological and evolutionary theories, based on methodological advances and an unprecedented availability of biodiversity big data; such epistemological and methodological innovations now support comprehensive predictive frameworks designed to formulate and test theoretically informed hypotheses on how human-induced effects change the most fundamental community properties, such as resilience and functional diversity (Raimundo, Guimarães Jr, & Evans, 2018).

Several works have applied the network approach to unravel the impacts of anthropogenic effects—such as fishing or hunting—on biodiversity, particularly regarding the consequences of species extinctions (Dunne et al., 2016; Gaichas & Francis, 2008; Navia et al., 2010, 2016; Pérez-Matus et al., 2017). Network modeling allows us to predict how species extinctions and other anthropogenic effects will affect ecological and evolutionary dynamics shaping food web structure affecting ecosystem functions (Ings, Montoya, Bascompte, et al., 2009; Raimundo, Marquitti, de Andreazzi, Pires, & Guimarães, 2018; Rohr & Bascompte, 2014; Thébault & Fontaine, 2010; Yen et al., 2016). Complementarily, structural analyses of ecological networks allow us to understand how patterns of interactions modulate demographic and evolutionary processes within communities, and how such dynamics change species-level properties, such as traits and abundances (Poisot, Stouffer, & Gravel, 2015). Structural analyses can also unravel topological roles that species play within networks, which inform how they influence the propagation of ecological and evolutionary effects (e.g., Olesen et al., 2007). Understanding and predicting how species-interaction networks are assembled and respond to anthropogenic disturbance is pivotal for the design of process-based strategies aimed to restore and conserve biodiversity and ecosystem functions at the community level (Raimundo, Guimarães Jr, & Evans, 2018).

2.2 *Macroscopic and Microscopic Patterns in Marine Food Webs: From Network Structure to Species Topological Roles*

Similar to species-rich terrestrial communities governed by joint demographic effects of multiple species interactions (Raimundo, Marquitti, et al., 2018), the structure and dynamics of marine food webs can be described and understood in terms of their topological patterns (Dunne et al., 2004). A growing body of evidence arising from structural analyses of trophic networks (e.g., Barabási, 2016; Lau, Borrett, Baiser, Gotelli, & Ellison, 2017; Navia, Cortés, Jordán, et al., 2012) supports that structural attributes are recurrent over different ecoregions. The major motivation to study macroscopic structure of ecological networks is that understanding how network architecture modulates the propagation of effects at the system level is key for the development of functional conservation strategies based on ecological and evolutionary processes that shape community structure and dynamics. Architectural patterns that are recurrent in ecological networks, such as modularity (Olesen et al., 2007; Rezende et al., 2009) and nestedness (Bascompte & Jordano, 2014; Cantor, Pires, Marquitti, Raimundo, et al., 2017), have been related to the stability of networks and the extent to which they can persist facing several types of environmental stressors (Bascompte et al., 2005; Dunne et al., 2004; Solé & Montoya, 2001; Thébault & Fontaine, 2010).

Structural analysis of marine food webs can also inform the roles that each species plays within an ecological community (Guimerà & Amaral, 2005; Jordán, Liu, & Davis, 2006; Olesen et al., 2007). Topological metrics that describe species-level properties within networks include species connectivity (*Degree*) and other indexes that inform species topological roles, for instance, regarding their centrality within network paths (*Betweenness, Closeness*) or their contribution to network cohesion (*network hubs*, *keystone index*, *topological uniqueness*) (Guimerà & Amaral, 2005; Jordán, 2001, 2009; Jordán et al., 2006; Jordán, Takacs-Santa, & Molnar, 1999; Krause, Frank, Mason, Ulanowicz, & Taylor, 2003; McMahon, Miller, & Drake, 2001). Understanding species topological roles enlightens how network properties emerge dynamically in the course of network evolution (Bornatowski et al., 2017; Gaichas & Francis, 2008). Theoretical and empirical research suggest that topological roles are driven by traits that mediate ecological interactions, such as niche breadth, individual motility, and trophic positions, alongside with community-specific properties that constrain patterns of interaction and species potential for interaction rewiring, such as phylogenetic clustering (Borthagaray et al., 2014; Guimerà et al., 2010; Olmo-Gilabert et al., 2019; Rezende et al., 2009).

Species topological roles represent a proxy for species contribution to community structure, which arguably can be extended to its dynamics. Importantly, we shall notice that topological roles are not static but change over time due to a variety of mechanisms driving species connectivity, which include temporal species turnover (Díaz-Castelazo, Sánchez-Galván, Guimarães Jr, Raimundo, & Rico-Gray, 2013) and several adaptive mechanisms of interaction rewiring (Raimundo, Guimarães Jr, & Evans, 2018; Valdovinos, Ramos-Jiliberto, Garay-Narvaez, Urbani, & Dunne, 2010). From an applied perspective, combining knowledge of macroscopic (global) network patterns and microscopic (local) information, such as provided by topological role analyses, can sustain testable predictions about system-level consequences of ongoing anthropogenic impacts affecting marine biodiversity (Bornatowski et al., 2014; Dambacher et al., 2010; Dunne et al., 2016; Navia et al., 2010, 2016, 2019).

Analyses of species topological roles represent the cornerstone for the development of innovative community-level conservation strategies because they allow the identification of species whose connectivity patterns lead to strong effects on intertwined demographic dynamics at the community level and, in addition, also represent key connections for the flow of matter and energy at the ecosystem level. For example, a few fish species, such as *Gadus macrocephalus* (Pacific cod) and *Hippoglossus stenolepis* (Pacific halibut), are structural connections that define the food web backbone in the Gulf of Alaska (Gaichas & Francis, 2008). In Brazil, large shark predators at higher trophic levels exert top-down effects that drive the ecological dynamics of species within lower trophic levels (Bornatowski et al., 2014). Species playing key topological roles are often those most connected, such as network hubs defining the backbone of modular networks that have many intra- and intermodular links (Olesen et al., 2007). Conversely, among-module synchronization in modular networks that miss network hubs will depend on connector species that link modules but are not necessarily highly connected (Guimerà & Amaral, 2005; Olesen et al., 2007). Additionally, under particular network architectures, such as small-world patterns and scale-free properties, species centrality does not correlate with structural patterns (Navia et al., 2010, 2016).

Marine food webs show recurrent topological patterns, such as small-world structures (Marina et al., 2018; Watts & Strogatz, 1998) and modularity (Newman & Girvan, 2004). Modularity has been reported for Arctic (Kortsch, Primicerio, Fossheim, et al., 2015), Antarctic (Saravia, Marina, De Troch, & Momo, 2018), temperate (Krause et al., 2003; Pérez-Matus et al., 2017), and tropical marine food webs (Rezende et al., 2009). Theory predicts that modularity will increase the stability in food webs (May, 1972, 1973; Pimm, 1979; Stouffer & Bascompte, 2011; Thébault & Fontaine, 2010) by preventing the spread of perturbations, i.e., modular structures would constrain the impact of disturbances either because the perturbations remain within a single compartment or reach other compartments with decreased magnitudes (Krause et al., 2003). Conversely, it has also been argued that under certain conditions, modularity can increase the likelihood of species extinctions and co-extinctions (Dáttilo, 2012) as deleterious effects remain within a module and will eventually be amplified by mechanisms such as feedback loops or extinction cascades. Despite its recurrent role as a major driver of marine biodiversity, a comprehensive overview of the applied consequences of modularity in marine food webs is mostly lacking.

2.3 Processes Shaping the Structure of Marine Food Webs

Several non-mutually exclusive mechanisms have been proposed to explain food web patterns. The first type of mechanisms refers to processes involving phenotypic traits that mediate ecological interactions (Cohen, Pimm, Yodzis, & Saldaña, 1993; Laigle et al., 2018), which are phylogenetically constrained (Cattin, Bersier, Banaek-Richter, Baltensperger, & Gabriel, 2004; Rezende et al., 2009; Webb,

Ackerly, McPeek, & Donoghue, 2002). The second class of mechanisms driving food web architecture encompasses neutral processes arising from spatio-temporal variation in abundance distributions, which determine the likelihood of pairwise species interactions based on encounter probabilities (Cohen, Jonsson, & Carpenter, 2003; Vázquez et al., 2007). Although the relative roles played by trait-based and neutral processes in shaping interaction networks vary over different biological systems, both contribute to network dynamics to some extent. Reciprocal effects between trait-based and neutral processes form feedback loops that connect ecological and evolutionary processes at the community level. Such feedback loops define eco-evolutionary dynamics that shape interaction patterns, abundance distributions, and trait diversity (see Raimundo, Guimarães Jr, & Evans, 2018 and the references therein).

We are still in the first steps regarding the application of eco-evolutionary theory to the management of biodiversity and ecosystem functions. Bridging knowledge on trait-interaction-abundance feedbacks that shape ecological network dynamics (Poisot et al., 2015) and the design of conservation strategies is key to ensure the much-needed inception of evolutionary community ecology into governance and policymaking (Jørgensen, Folke, & Carroll, 2019). Facing such a challenge, we shall consider that abundances and traits can vary widely over space and time and, in addition, are influenced by processes at different levels of biological organization: from genes to individuals, populations, communities, and ecosystems (Pacheco, Traulsen, & Nowak, 2006). In order to integrate several classes of structuring processes into robust predictive frameworks accounting for the dynamics of marine biodiversity, we can rely on dynamic network models that have been increasingly applied to investigate the interplay among ecological and evolutionary processes that account for community properties (e.g., Poisot et al., 2015; Suweis et al., 2013; Zhang, Hui, & Terblanche, 2011).

2.4 *The Roles That Fisheries Play Within Food Webs*

The development of dynamic network models that can enlighten fishing and conservation policies requires a proper consideration of the interplay between anthropogenic and eco-evolutionary processes shaping contemporary marine biodiversity. Human-induced changes in marine ecosystems have occurred for centuries but only in the past decades their impacts reached a global scale (Jackson et al., 2001; Lotze et al., 2011; McCauley et al., 2015). There are three major anthropogenic effects that affect marine biodiversity and ecosystem functioning, namely: (1) changes in nutrient cycles and climate, which affect bottom-up ecosystem processes, (2) fishing activities, which impose top-down effects threatening species diversity, and (3) habitat degradation and contamination, which affect processes across all trophic levels (Navia, Cortés, Jordán, et al., 2012). Since fisheries impose both direct and indirect effects on marine ecosystems, fishing impacts on target and non-target fish species are likely to propagate within the food web and emerge as changes in ecological properties at the community level (Standström et al., 2005).

To date, most studies have focused on the effects of fishing on population dynamics, most often of charismatic species or of taxa with higher commercial values (e.g., Lotze, 2009; Lotze et al., 2011). Although some works have assessed the effects of anthropogenic activities on functional groups or entire food webs (e.g., Jackson et al., 2001; Pandolfi, Bradbury, Sala, et al., 2003), the consequences of human impacts on the structure and functioning of marine ecosystems remain mostly unclear (Lotze et al., 2011). On the one hand, several studies found that fisheries reshape patterns of species interactions whose effects propagate at the community level and change the distributions of species abundances (Myers, Hutchings, & Barrowman, 1996) and of traits that mediate ecological interactions (Bianchi, Gislason, Graham, et al., 2000; Jennings et al., 1999), which, in turn, trigger further changes in network organization and dynamics (Barraclough, 2015; Myers, Baum, Shepherd, Powers, & Peterson, 2007; Poisot et al., 2015; Raimundo, Guimarães Jr, & Evans, 2018). Therefore, it can be generally expected that the consequences of fishery-induced extinctions will reshape network architecture (Dunne et al., 2016; Pérez-Matus et al., 2017), community composition (Hutchings & Baum, 2005), and trophic structure (Ferretti, Worm, Britten, Heithaus, & Lotze, 2010; Pauly et al., 1998), such as have been reported for temperate (Coll, Palomera, & Tudela, 2009; Gaichas & Francis, 2008; Myers et al., 2007; Shepherd & Myers, 2005) and tropical marine ecosystems (Navia & Mejía-Falla, 2016; Stevens, Bonfil, Dulvy, & Walker, 2000).

Fisheries often act as adaptive foragers within ecosystems by optimizing capture rates (Begossi, 1992; Bertrand, Bertrand, Guevara-Carrasco, & Gerlotto, 2007; Poos & Rijnsdorp, 2007) and by switching target species to minimize costs and maximize profits (Acheson, 1988; Sethi, Branch, & Watson, 2010). Such dynamic fishing patterns are expected to trigger eco-evolutionary feedbacks that will propagate over the whole community and reshape abundance and trait distributions (Barraclough, 2015; Tromeur & Loeuille, 2018). The consequences of adaptive foraging by fisheries at the ecosystem level remain unclear. Adaptive harvesting can arguably stabilize marine ecological communities by continuously reallocating fishing pressures to target more abundant prey species, thereby counterbalancing negative effects of fishing on rare species, releasing them from interspecific competition and hence contributing to community stability and diversity (Kondoh, 2003; Loeuille, 2010). On the other hand, adaptive interaction switches performed by fisheries could induce abrupt phase shifts with unexpected consequences for community structure and dynamic (Conversi, Dakos, Gårdmark, Ling, et al., 2015; Estes et al., 2016; Estes, Terborgh, Brashares, et al., 2011; Jackson et al., 2001). To shed light on such contradictory predictions regarding the network-level consequences of fishing impacts, fisheries can be incorporated into food web analyses as additional nodes that are analogous to high-level predators. Incorporating fisheries as dynamic agents within the food webs, whose interaction patterns shape and are shaped by community-level processes, can greatly improve our ability to understand, predict, and manage fishing impacts (Dunne et al., 2016; Maschner et al., 2009; Pérez-Matus et al., 2017). It is widely acknowledged by ecologists that several types of bottom-up and top-down effects imposed by single keystone species can lead to broad changes at the community level (Jones, Lawton, & Shachak, 1994; Paine, 1969). Similarly, fisheries are likely to act as network hubs (Guimerà & Amaral, 2005; Olesen et al., 2007) and hence drive marine food web structure and dynamics.

3 Analytical Approaches

Topological changes in food webs arise as a consequence of the adaptive rewiring of multi-species interactions due to a variety of ecological and evolutionary mechanisms (see Raimundo, Guimarães Jr, & Evans, 2018). Understanding the extent to which such structural changes in trophic networks affect ecosystem dynamics and stability is a key issue for biodiversity conservation (de Ruiter, Wolters, Moore, & Winemiller, 2005; Pimm, 2002). The perception that community- and ecosystem-level approaches are much-needed to conserve biodiversity led to integrative conceptual frameworks, such as the multispecific management (May, Beddington, Clark, Holt, & Laws, 1979; Yodzis, 2000) and ecosystem approaches (Grant, Marin, & Pedersen, 1997), which recognize the need to take into account not only functional roles played by single species but consider responses of whole species assemblages to more accurately model food web dynamics (Jordán et al., 2006). Although changes in species composition can be used as a proxy for ecosystem perturbations, a systemic approach capable of providing comprehensive indicators of changes in structural and functional properties of ecological communities depends on the description of species-interaction networks, which account to the ecological and evolutionary processes driving biodiversity dynamics at the community level (Bascompte et al., 2005; Dunne et al., 2002).

Natural trophic networks show recurrent structural properties across ecosystems, which provide benchmarks for the evaluation of biodiversity erosion within degraded ecosystems subject to intense human activities (Dell et al., 2005). Network analyses provide several tools that can connect theoretical foundations and methodological tools provided by quantitative community ecology to food web conservation and management. In what follows, we summarize analytical approaches commonly used to describe food web properties and discuss their current and potential interface with marine biodiversity conservation and management.

3.1 *Structural Analysis: Local and Mesoscale Indices*

Several local and mesoscale network indices, most of which consider distances between nodes (Wasserman & Faust, 1994), are widely applied to describe network- or species-level properties within marine food webs. Such indices take into account every possible (direct and indirect) interaction between species *i* and *j*. Regarding local indices, the most fundamental metric describing a species (node) is its degree (*D*), which is computed as the number of direct links to other species (nodes) for both prey (in-degree) and predators (out-degree) (Jordán et al., 2006). If the degrees of all nodes within the network are known, one can describe the distribution of links, i.e., the statistical distribution of degree-values that can readily inform key network-level properties. For example, if the degree distribution follows a power law, the network is said to be scale-free (Dunne et al., 2002; Montoya & Solé, 2002; Solé & Montoya, 2001).

Several topological indices are built on information about node neighborhood and interaction paths between neighbors, such as a family of well-known metrics named "Centrality indices." For example, betweenness centrality (*BC*) measures how a node is incident to many shortest paths in the network. If a trophic group *i* has a high BC_i value, then removing this group from the network decreases the overall degree of network synchronization, as interaction paths accounting for rapidly spreading effects will no longer be available (Jordán et al., 2006). Closeness centrality (*CC*) is another widely used metric from this family of indices, which quantifies the length of minimal paths from any given node to all others (Wasserman & Faust, 1994). Removing trophic groups characterized by high CC_i will have repercussions on most other groups within the network (Jordán et al., 2006).

Given the structural complexity of species-rich interaction networks, the graphic representation of indices such as those described above is an important tool to assist the interpretation of the results, especially when the aim is to communicate findings to non-scientific audiences or decision-makers (Pocock et al., 2016). As an example, we present graphs resulting from centrality analyses of a binary matrix (Navia, 2013, Fig. 1a) depicting predator–prey interactions at the Gulf of Tortugas, Colombian Pacific coast. First, we plotted the overall network structure by showing undirected interactions between predators and prey (Fig. 1b). Subsequently, we computed degree (*D*), betweenness centrality (*BC*), and closeness centrality (*CC*) to gain insight into key local and global network properties. In Fig. 1c, node sizes are proportional to their degrees and hence it is possible to identify the most connected species (nodes), such as shrimps (green nodes), sharks (gray nodes), and rays (black nodes). Based on betweenness centrality (*BC*), we detected that some species of bony fishes (red nodes), sharks and rays (black and gray nodes), and shrimps (green nodes) are likely to drive the propagation of ecological and evolutionary effects within the network (Fig. 1d). Finally, closeness centrality (*CC*, smaller nodes correspond to higher closeness values) informs that predators, such as sharks and rays, alongside with the most consumed prey, such as shrimp and squids, are key drivers of effect propagation within the food web (Fig. 1e). These examples illustrate how the structural analyses of ecological networks can rapidly provide information for fishing management measures, which in this case should focus on shrimp, sharks, and rays that are key drivers of network dynamics. Comprehensive information on the above-mentioned indices and their application to marine environments are provided by Wasserman and Faust (1994), Jordán et al. (2006), Abarca-Arenas, Franco-López, Peterson, Brown-Peterson, and Valero-Pacheco (2007), Dambacher et al. (2010), Navia et al. (2010, 2016, 2019), Oshima and Leaf (2018).

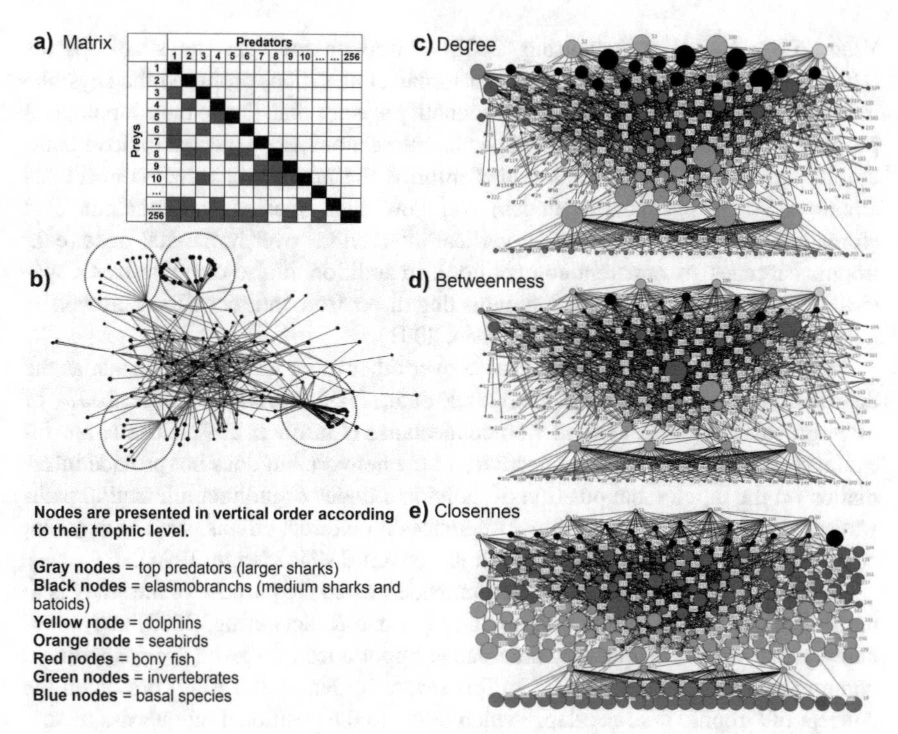

Fig. 1 Visualization of degrees and centrality indices of food web of the Gulf of Tortugas, Colombian Pacific coast (Figures c, d, e are modified from Navia, 2013)

3.2 Keystone Species and Keystone Species Complexes

Many of the indexes mentioned so far consider interactions between neighboring nodes, but not necessarily whole systems (Benedek, Jordán, & Báldi, 2007). Species can be in pivotal topological positions within ecological networks, exerting disproportionately strong structural and functional effects at the community level (Capocefalo, Pereira, Mazza, & Jordán, 2018). The propagation of secondary effects (e.g., trophic cascades or apparent competition; Menge, 1995) within food webs is hard to predict without high-quality information on network structure. Fortunately, the increasing availability of species-interaction big data now provides unprecedented conditions for ecologists to apply structural network analysis in order to identify species that play key topological roles and properly sustain system-level biodiversity conservation strategies (Capocefalo et al., 2018; Raimundo, Guimarães Jr, & Evans, 2018).

The use of quantitative species-interaction data to detect those taxa that act as key network nodes (defined as topological keystone species, Jordán et al., 1999, 2006; Libralato, Christensen, & Pauly, 2006) can greatly improve system-level conservation strategies because the extinction of such species will cause stronger effects at the network level (Allesina, Bodini, & Bondavalli, 2006; Jordán, Liu, & van

Veen, 2003; Jordán & Scheuring, 2002). Based on the "net status" of species (Harary, 1961), Jordán et al. (1999) and Jordán et al. (2006) proposed the keystone index as a straightforward procedure to identify species that, due to their topological position within the network of interspecific interactions, are expected to drive community dynamics. The keystone index informs the number of direct connections among neighboring species (nodes) and how such species are interconnected (Jordán et al., 2006), emphasizing vertical interactions over horizontal ones (e.g., trophic cascades vs. apparent competition). In addition, it also characterizes a species' positional importance by distinguishing direct from indirect effects, as well as bottom-up from top-down effects (Jordán, 2001).

Mesoscale indices have been favored over other more local indices, such as the distribution of trophic connections (Dunne et al., 2002; Montoya & Solé, 2002), or more global ones, such as food web connectance (Martinez, 1992). The latter, for example, reflects the global connectivity of the network but does not provide information on the topological position of individual nodes or indirect interaction pathways and, therefore, does not allow inferences on indirect effects, such as apparent competition and trophic cascades (Holt & Lawton, 1994; Menge, 1995).

Overall, mesoscale indices are recommended when the purpose of the study is to unravel relationships within a community (Jordán & Scheuring, 2002) and, especially, when we aim to quantify the relative importance of a given species within a community (Jordán et al., 2006). In this sense, Jordán et al. (2009) presented the concept of "trophic field overlap," which refers to the positional uniqueness of species. The corresponding overlap metric quantifies species topological redundancy within interaction networks and hence identifies species showing rich, as well as unique interaction patterns (Jordán et al., 2009). Using the indices of topological importance and redundancy, Navia et al. (2016) found that ecosystem-based fisheries management should prioritize not only highly central species (e.g., shrimps, which are species of high commercial value in the area, white nodes 170, 171, 173, and 174, Fig. 2) but should also consider species with unique structural properties such as sharks with low topological redundancy (black nodes with numbers 1 until 4; Fig. 2). The topological keystone index (KI) and topological uniqueness (TU) can also be graphed to facilitate the interpretation of the results (Fig. 2). In this example, the values of TU with and without top predators depict the structural effect that the loss of these species would have on the network and how the topological redundancy would be affected. When the extinction of top predators (large sharks) is simulated, the lower values of topological redundancy are transferred to bony fishes (gray nodes 1 and 3), marine mammals (yellow node), and to medium-sized shark and ray species, implying the reorganization of the top-down structure of the network.

Nevertheless, beyond using information on the positional importance of single nodes within networks, the development of system-level conservation strategies can benefit from analyses that inform (1) the topological roles of whole sets of species (nodes) or, complementarily, (2) exactly which set of nodes is the most important in maintaining network integrity (by quantifying the structural effect of its deletion). Network analyses considering multi-species sets are founded on both empirical

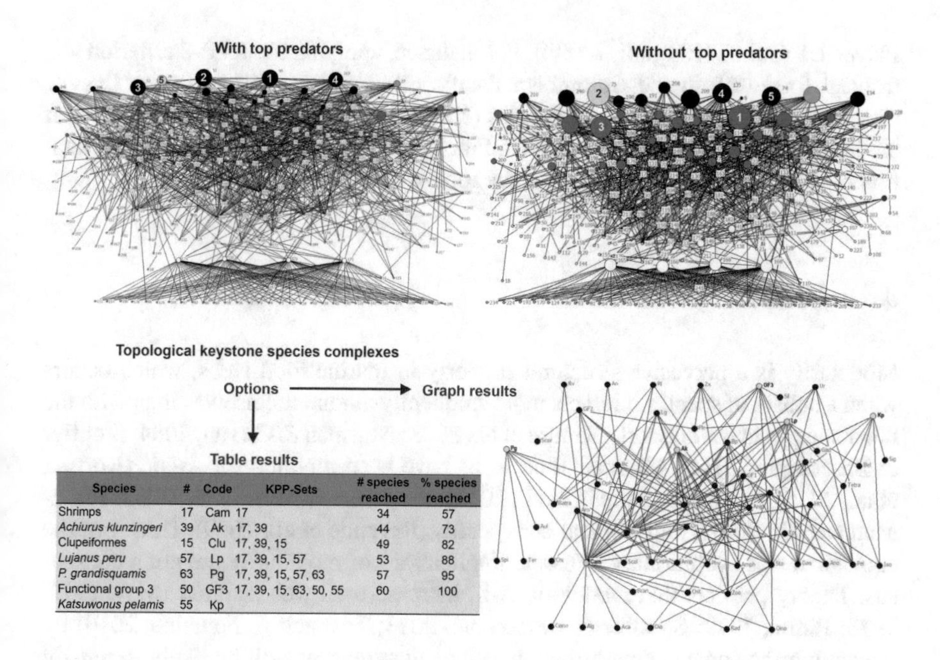

Species	#	Code	KPP-Sets	# species reached	% species reached
Shrimps	17	Cam	17	34	57
Achiurus klunzingeri	39	Ak	17, 39	44	73
Clupeiformes	15	Clu	17, 39, 15	49	82
Lujanus peru	57	Lp	17, 39, 15, 57	53	88
P. grandisquamis	63	Pg	17, 39, 15, 57, 63	57	95
Functional group 3	50	GF3	17, 39, 15, 63, 50, 55	60	100
Katsuwonus pelamis	55	Kp			

Fig. 2 Visualization of topological uniqueness (TU) with and without top predators and options to show results of the Key Player problem (table or figure). See the text for methodological details. (Figure with top predators taken and modified from Navia et al., 2016)

knowledge and modeling (Daily, Ehrlich, & Haddad, 1993; Ortiz et al., 2013, 2015, 2017), which together allow the identification of keystone species complexes that should be prioritized in conservation strategies (Daily et al., 1993). Indeed, current studies in landscape ecology (Pereira & Jordán, 2017; Pereira, Saura, & Jordán, 2017) reinforce the notion that the positional importance of species within food webs should be characterized simultaneously rather than independently in order to make system-level conservation strategies more effective.

Despite the potential of analyzing keystone species complexes as a tool to guide the design of marine biodiversity conservation strategies, few works have successfully addressed this issue. A promising example is the application of the concept of key species groups (Daily et al., 1993) to solve the Key Player (KP) problem (Borgatti, 2003a) in the context of food webs. According to the proposal of Benedek et al. (2007) and Ortiz et al. (2013), based on quantitative or semi-quantitative models (mass-balance and *loop analysis*), keystone species complexes (KSCs) consist of an interaction core formed by species and/or functional groups linked by strong interspecific interactions. The next step is to identify which KSCs maximize network integrity according to two criteria. First, "Fragmentation" (KPP-1) is computed as the contribution of each node to network cohesion and informs the identity of nodes whose removal will maximize network fragmentation. The second criterion, "Expansion" (KPP-2), identifies which KSCs maximize the fast propagation of effects over the whole network. This analysis can be performed using the Key

Player 1.1 routine (Borgatti, 2003b). For instance, applying the KPP-2 criterion to a tropical food web in the Colombian Pacific coast without top predators (Navia, unpublished data) shows that six species (KPP-2 sets)—including mostly species of medium trophic levels (yellow nodes in Fig. 2)—are required to ensure the propagation of ecological effects to 100% of the food web.

3.3 Modularity

Modularity is a pervasive structural property in marine food webs, which occurs when a subset of species interacts more frequently among themselves than with the other species in the network (Krause et al., 2003; Newman & Girvan, 2004; Stouffer & Bascompte, 2011). Modular food webs have been reported for Arctic (Kortsch et al., 2015), Antarctic (Saravia et al., 2018), temperate (Krause et al., 2003; Pérez-Matus et al., 2017), and tropical ecosystems (Rezende et al., 2009). However, the ecological and evolutionary dynamic implications of modularity are not a consensus. Theory predicts that modularity will increase food web stability (May, 1972, 1973; Pimm, 1979; Stouffer & Bascompte, 2011; Thébault & Fontaine, 2010) by preventing the spread of perturbations, since disturbances will be likely to remain within a single compartment or reach other compartments with decreased magnitudes (Krause et al., 2003).

For modular food webs, the topological roles that species play result from the balance between their connectivity between and within modules (Guimerà & Amaral, 2005; Saravia et al., 2018). Such a balance implies four categories of topological roles: (1) network hubs, which are species showing high degrees of both intra- and intermodular connectivity, i.e., these species form the network backbone and hence their removal is expected to have broad structural and dynamical consequences; (2) module hubs, which are species that show low intermodular connectivity but high intramodular connectivity, i.e., these species drive local dynamics; (3) connectors, which are species showing high intermodular connectivity but low intramodular connectivity, i.e., these species do not belong to modules but can be unique connections between large modules, even if they have a low number of connections; (4) peripherals, which are species with low intra- and intermodular connectivity and hence expected to have minimal impacts on community dynamics (Guimerà & Amaral, 2005; Olesen et al., 2007).

A preliminary analysis of topological roles in a modular tropical marine food web (Fig. 3; Márquez-Velásquez et al., unpublished data) shows that a small subset of species acting as network hubs (2% of the species pool) and module hubs (4% of the species pool) account for the network backbone. Among these, three shark species act as network hubs, suggesting that top-down effects drive food web dynamics. Shrimp species, some bony fish species as Pacific snappers and catfishes, and a stingray act as module hubs and hence also are pivotal elements sustaining network compartmentalization. Up to 85% of the species pool is formed by peripheral species, which are only locally connected. Species playing different topological roles are dispersed over all food web modules, which might suggest that species in the same module share functional traits, such as body mass (Laigle et al., 2018). These results support that multi-species management should go beyond protecting top predators and include at least module hubs and connectors to ensure the long-term persistence of community structure.

3.4 Linking Network Theory and Modeling: An Adaptive Network Models Approach

Over the last two decades, empirical knowledge on ecological network structure (e.g., Pascual & Dunne, 2006; Rezende et al., 2009) associated with the development of a diversity of structural analyses (e.g., Marina et al., 2018; Newman & Girvan, 2004) and network modeling approaches (e.g., Allesina & Tang, 2015; Gross & Blasius, 2008) has set exceptional conditions for the design of novel approaches linking theoretical and applied ecology and aimed to improve our understanding of how species interactions modulate ecological and evolutionary dynamics at the whole-community level (e.g., Thébault & Fontaine, 2010).

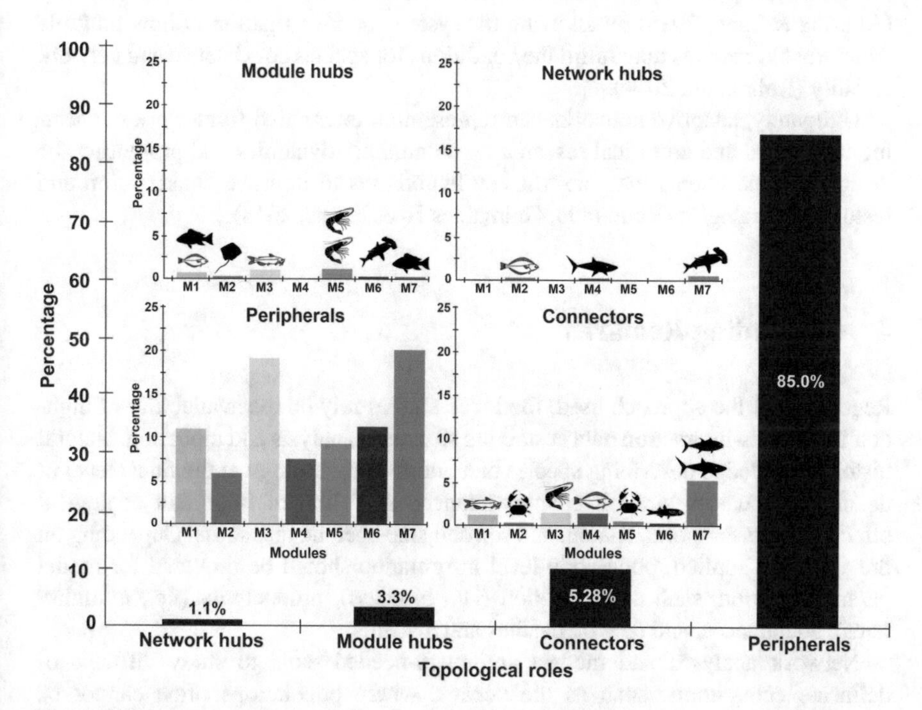

Fig. 3 Illustrative modular structure and topological roles of a tropical marine food web (Márquez-Velásquez et al., unpublished data)

Dynamic network modeling combined with computer simulations can provide theoretically founded predictions of how biodiversity loss is expected to reshape community-level properties, such as resilience and functional diversity, and influence the long-term persistence of ecosystems (e.g., Thébault & Fontaine, 2010; Vinagre, Costa, Wood, Williams, & Dunne, 2019; Yen et al., 2016). Particularly, adaptive networks are a promising conceptual and modeling framework (Gross & Blasius, 2008; Gross & Sayama, 2009) that can help us to address trait-interaction-abundance feedbacks (Poisot et al., 2015) and deepen our understanding of eco-evolutionary mechanisms driving long-term community dynamics. An adaptive network is defined by the feedback loop between network structure and the properties of the nodes, i.e., it encompasses the mutual effects between changes in interaction patterns and associated network properties that characterize community structure and dynamics, such as modularity and stability, and population-level, eco-evolutionary processes shaping the properties of species that form the network, such as their abundances and trait values (Gross & Blasius, 2008; Gross & Sayama, 2009).

Dynamic network models incorporating adaptive rewiring of interactions can help us to predict, for example, how improved fisheries regulations may promote competition releases and change the interaction patterns of natural network hubs, which may have broader structural consequences for network structure and ecological dynamics (Valdovinos et al., 2010). This integrative framework can shed light on the relationship between ecological network structure and ecosystem stability (Allesina & Tang, 2015) by allowing the systematic investigation of how multiple structural alternatives may fulfill the conditions for species coexistence and network stability (Rohr et al., 2014).

Ultimately, adaptive networks can represent an integrative framework connecting theoretical and empirical research on community dynamics and promoting the design of experiments that can test key hypotheses to improve conservation and restoration strategies (Raimundo, Guimarães Jr, & Evans, 2018).

4 Concluding Remarks

Regardless of the approach used, food web studies rely on the availability of high-quality species-interaction data sustaining structural analysis and modeling. Natural history knowledge describing species diet and feeding ecology at the finest level of detail is mandatory since it enables nuanced modeling of important ecological effects such as temporal, spatial, and sex and size-specific diet shifts. Depending on the approach applied, population-level information should be provided for model parameterization, such as production (i.e., biomass), productivity (i.e., mortality rates), abundances, and data on catches and discards.

Network analyses and models are much-needed tools to study difficult-to-delineate ecosystems, such as the oceans, where populations often cannot be manipulated and hence inferences on the relationship between species loss and community stability are not easy to obtain using empirical approaches. The network ecologist toolbox can partially reproduce marine ecosystems complexity and perform "computer experiments" that simulate community dynamics under several perturbation scenarios and conservation strategies. Predictions provided by network models can provide system-level information on biodiversity and ecosystem functions to decision-makers, which would otherwise be very difficult to obtain. Conversely, modelers should not neglect the need for empirically testing their assumptions and predictions to avoid indiscriminate errors of extrapolation or overreaching conclusions.

Advances in ecological network modeling can allow us to understand and predict natural and anthropogenic impacts on marine biodiversity and the consequences of species extinctions. Therefore, understanding how interaction among marine species is assembled and responds to anthropogenic disturbance is pivotal for the design of process-based strategies aimed to conserve marine biodiversity at the community level. Species topological roles are not static but change over time due to a variety of mechanisms driving species connectivity, including temporal species turnover (Díaz-Castelazo et al., 2013) and adaptive mechanisms of interaction rewiring (Raimundo, Guimarães Jr, & Evans, 2018). The next step to improve the application of the network approach to understand and manage marine biodiversity and ecosystem functions is to investigate topological roles of species change under anthropogenic pressures and its dynamic impacts on the systems.

A broad challenge that we shall face in order to advance the network approach to understand and manage marine biodiversity is to bridge dynamic network modeling, species-interaction big data, and biomonitoring approaches by using parameters that can be easily estimated in the field by conservation practitioners. Facing such a challenge, we can benefit from the ongoing theoretical synthesis in ecology and evolution and a variety of novel methods for network analyses and modeling. This exciting perspective may allow us to combine theoretical and empirical community ecology to shed light on the processes driving ecological networks and improve our ability to manage marine biodiversity and ecosystem functions.

References

Abarca-Arenas, L. G., Franco-López, J., Peterson, M. S., Brown-Peterson, N. J., & Valero-Pacheco, E. (2007). Sociometric analysis of the role of penaeids in the continental shelf food web off Veracruz, Mexico based on by-catch. *Fisheries Research, 87*(1), 46–57.

Acheson, J. M. (1988). Patterns of gear changes in the Marine fishing industry: Some implications for management. *Marine Anthropological Studies, 1*, 49–65.

Allesina, S., Bodini, A., & Bondavalli, C. (2006). Secondary extinctions in ecological networks: Bottlenecks unveiled. *Ecological Modelling, 194*(1-3), 150–161.

Allesina, S., & Tang, S. (2015). The stability–complexity relationship at age 40: A random matrix perspective. *Population Ecology, 57*(1), 63–75.

Barabási, A. L. (2016). *Network science*. Cambridge University Press.

Barraclough, T. G. (2015). How do species interactions affect evolutionary dynamics across whole communities? *Annual Review of Ecology, Evolution, and Systematics, 46*, 25–48.

Bascompte, J., & Jordano, P.. (2014). *Mutualistic networks*. Monographs in population biology series, no. 53. Princeton University Press, Princeton, 216 pp.

Bascompte, J., Melián, C. J., & Sala, E. (2005). Interaction strength combinations and the overfishing of a marine food web. *Proceedings of the National Academy of Sciences of the United States of America, 102*(15), 5443–5447.

Baum, J. K., & Worm, B. (2009). Cascading top-down effects of changing oceanic predator abundances. *The Journal of Animal Ecology, 78*(4), 699–714.

Bax, N., Williamson, A., Aguero, M., Gonzalez, E., & Geeves, W. (2003). Marine invasive alien species: A threat to global biodiversity. *Marine Policy, 27*(4), 313–323.

Begossi, A. (1992). The use of optimal foraging theory in the understanding of fishing strategies: A case from Sepetiba Bay (Rio de Janeiro State, Brazil). *Human Ecology, 20*, 463–475.

Benedek, Z., Jordán, F., & Báldi, A. (2007). Topological keystone species complexes in ecological interaction networks. *Community Ecology, 8*, 1–7.

Bertrand, S., Bertrand, A., Guevara-Carrasco, R., & Gerlotto, F. (2007). Scale-invariant movements of fishermen: The same foraging strategy as natural predators. *Ecological Applications, 17*, 331–337.

Bianchi, G., Gislason, H., Graham, K., Hill, L., Jin, X., Koranteng, K., et al. (2000). Impact of fishing on size composition and diversity of demersal fish communities. *ICES Journal of Marine Science, 57*(3), 558–571.

Borgatti, S. P. (2003a). The key player problem. In R. Breiger, K. Carley, & P. Pattison (Eds.), *Dynamic social network modeling and analysis: Workshop summary and papers* (pp. 241–252). Washington DC: Committee on Human Factors, The National Academies Press, 381 pp.

Borgatti, S. P. (2003b). Key player. Analytic Technologies. http://www.analytictech.com/

Bornatowski, H., Navia, A. F., & Barreto, R. R. (2017). Topological redundancy and "small-world" patterns in a food web in a subtropical ecosystem of Brazil. *Marine Ecology, 38*, e12407.

Bornatowski, H., Navia, A. F., Braga, R. R., Abilhoa, V., & Corrêa, M. F. M. (2014). Ecological importance of sharks and rays in a structural food web analysis in southern Brazil. *ICES Journal of Marine Science, 71*(7), 1586–1592.

Borthagaray, A. I., Arim, M., & Marquet, P. A. (2014). Inferring species roles in metacommunity structure from species co-occurrence networks. *Proceedings of the Biological Sciences, 281*, 2014–2025.

Cantor, M., Pires, M. M., Marquitti, F. M., Raimundo, R. L., Sebastián-González, E., Coltri, P. P., et al. (2017). Nestedness across biological scales. *PLoS One, 12*(2).

Capocefalo, D., Pereira, J., Mazza, T., & Jordán, F. (2018). Food web topology and nested keystone species complexes. Complexity 2018.

Cattin, M. F., Bersier, L. F., Banaek-Richter, C., Baltensperger, R., & Gabriel, J. P. (2004). Phylogenetic constraints and adaptation explain food-web structure. *Nature, 427*(6977), 835.

Cohen, J. E., Jonsson, T., & Carpenter, S. R. (2003). Ecological community description using the food web, species abundance, and body size. *Proceedings of the National Academy of Sciences of the United States of America, 100*, 1781–1786.

Cohen, J. E., Pimm, S., Yodzis, P., & Saldaña, J. (1993). Body sizes of animal predators and animal prey in food webs. *The Journal of Animal Ecology, 62*, 67–78.

Coll, M., Palomera, I., & Tudela, S. (2009). Decadal changes in a NW Mediterranean Sea food web relation to fishing exploitation. *Ecological Modelling, 220*, 2088–2102.

Conversi, A., Dakos, V., Gårdmark, A., Ling, S., Folke, C., Mumby, P. J., et al. (2015). A holistic view of marine regime shifts. *Philosophical Transactions of the Royal Society B: Biological Sciences, 370*(1659), 20130279.

Daily, G. C., Ehrlich, P. R., & Haddad, N. M. (1993). Double keystone bird in a keystone species complex. *Proceedings of the National Academy of Sciences of the United States of America, 90*, 592–594.

Dambacher, J. M., Young, J. W., Olson, R. J., Allain, V., Galvan-Magana, F., Lansdell, M. J., et al. (2010). Analyzing pelagic food webs leading to top predators in the Pacific Ocean: A graph-theoretic approach. *Progress in Oceanography, 86*, 152–165.

Dáttilo, W. (2012). Different tolerances of symbiotic and nonsymbiotic ant-plant networks to species extinctions. *Network Biology, 2*(4), 127.

de Ruiter, P. C., Wolters, V., Moore, J. C., & Winemiller, K. O. (2005). Food web ecology: Playing Jenga and beyond. *Science, 309*, 68–71.

Dell, A. I., Kokkoris, G. D., Banasek-Richter, C., Bersier, L. F., Dunne, J. A., Kondoh, M., et al. (2005). How do complex food webs persist in nature? In M. P. C. de Ruiter, V. Wolters, & J. C. Moore (Eds.), *Dynamic food webs: Multispecies assemblages, ecosystem development and environmental change* (pp. 425–436). San Diego, CA: Academic Press. isbn:978-0-12-088458-2.

Díaz-Castelazo, C., Sánchez-Galván, I. R., Guimarães, P. R., Jr., Raimundo, R. L. G., & Rico-Gray, V. (2013). Long-term temporal variation in the organization of an ant–plant network. *Annals of Botany, 111*(6), 1285–1293.

Dunne, J. A., Maschner, H., Betts, M. W., Huntly, N., Russell, R., Williams, R. J., et al. (2016). The roles and impacts of human hunter-gatherers in North Pacific marine food webs. *Scientific Reports, 6*, 21179.

Dunne, J. A., Williams, R. J., & Martinez, N. D. (2002). Network structure and biodiversity loss in food webs: Robustness increases with connectance. *Ecology Letters, 5*, 558–567.

Dunne, J. A., Williams, R. J., & Martinez, N. D. (2004). Network structure and robustness of marine food webs. *Marine Ecology Progress Series, 273*, 291–302.

Endrédi, A., Jordán, F., & Abonyi, A. (2018). Trait-based paradise–or only feeding the computer with biology? *Community Ecology, 19*(3), 319–321.

Essington, T. E., Beaudreau, A. H., & Wiedenmann, J. (2006). Fishing through marine food webs. *Proceedings of the National Academy of Sciences, 103*(9), 3171–3175.

Estes, J. A., Heithaus, M., McCauley, D. J., Rasher, D. B., & Worm, B. (2016). Megafaunal impacts on structure and function of ocean ecosystems. *Annual Review of Environment and Resources, 41*, 83–116.

Estes, J. A., Terborgh, J., Brashares, J. S., Power, M. E., Berger, J., Bond, W. J., et al. (2011). Trophic downgrading of planet earth. *Science, 333*, 301–306.

Ferretti, F., Worm, B., Britten, G. L., Heithaus, M. R., & Lotze, H. K. (2010). Patterns and ecosystem consequences of shark declines in the ocean. *Ecology Letters, 13*(8), 1055–1071.

Gaichas, S. K., & Francis, R. C. (2008). Network models for ecosystem-based fishery analysis: A review of concepts and application to the Gulf of Alaska marine food web. *Canadian Journal of Fisheries and Aquatic Sciences, 65*(9), 1965–1982.

Gilljam, D., Curtsdotter, A., & Ebenman, B. (2015). Adaptive rewiring aggravates the effects of species loss in ecosystems. *Nature Communications, 6*, 8412.

Grant, W., Marin, S. L., & Pedersen, E. K. (1997). *Ecology and natural resource management: Systems analysis and simulation*. New Jersey: Wiley.

Gross, T., & Blasius, B. (2008). Adaptive coevolutionary networks: A review. *Journal of the Royal Society Interface, 5*, 259–271.

Gross, T., & Sayama, H. (2009). Adaptive networks. In *Adaptive networks* (pp. 1–8). Berlin: Springer.

Guimarães, P. R., Jr., Pires, M. M., Jordano, P., Bascompte, J., & Thompson, J. N. (2017). Indirect effects drive coevolution in mutualistic networks. *Nature, 550*(7677), 511–514.

Guimerà, R., & Amaral, L. A. N. (2005). Cartography of complex networks: Modules and universal roles. *Journal of Statistical Mechanics Theory and Experiment, 2*, P02001.

Guimerà, R., Stouffer, D. B., Sales-Pardo, M., Leicht, E. A., Newman, M. E. J., & Amaral, L. A. N. (2010). Origin of compartmentalization in food webs. *Ecology, 91*, 2941–2951.

Harary, F. (1961). Who eats whom. *General Systems, 6*, 41–44.

Hillebrand, H., Brey, T., Gutt, J., Hagen, W., Metfies, K., Meyer, B., et al. (2018). Climate change: Warming impacts on marine biodiversity. In *Handbook on marine environment protection* (pp. 353–373). Cham, Switzerland: Springer.

Hoegh-Guldberg, O., & Bruno, J. F. (2010). The impact of climate change on the world's marine ecosystems. *Science, 328*, 1523–1528.

Holt, R. D., & Lawton, J. (1994). The ecological consequences of shared natural enemies. *Annual Review of Ecology and Systematics, 25*(1), 495–520.

Hutchings, J. A., & Baum, J. K. (2005). Measuring marine fish biodiversity: Temporal changes in abundance, life history and demography. *Philosophical Transactions of the Royal Society of London. Series B, Biological Sciences, 360*(1454), 315–338.

Ings, T. C., Montoya, J. M., Bascompte, J., Blüthgen, N., Brown, L., Dormann, C. F., et al. (2009). Ecological networks–beyond food webs. *The Journal of Animal Ecology, 78*(1), 253–269.

Islam, M. S., & Tanaka, M. (2004). Impacts of pollution on coastal and marine ecosystems including coastal and marine fisheries and approach for management: A review and synthesis. *Marine Pollution Bulletin, 48*(7–8), 624–649.

Jackson, J. B., Kirby, M. X., Berger, W. H., Bjorndal, K. A., Botsford, L. W., Bourque, B. J., et al. (2001). Historical overfishing and the recent collapse of coastal ecosystems. *Science, 293*, 629–637.

Jennings, S., Greenstreet, S. P., & Reynolds, J. D. (1999). Structural change in an exploited fish community: A consequence of differential fishing effects on species with contrasting life histories. *The Journal of Animal Ecology, 68*(3), 617–627.

Jennings, S., & Kaiser, M. J. (1998). The effects of fishing on marine ecosystems. *Advances in Marine Biology, 34*, 201–350.

Jones, C. G., Lawton, J. H., & Shachak, M. (1994). Organisms as ecosystem engineers. *Oikos, 763*(69), 373–386.

Jordán, F. (2001). Strong threads and weak chains? - A graph theoretical estimation of the power of indirect effects. *Community Ecology, 2*, 17–20.

Jordán, F. (2009). Keystone species and food webs. *Philosophical Transactions of the Royal Society B Biological Sciences, 364*(1524), 1733–1741.

Jordán, F., Liu, W. C., & Davis, A. (2006). Topological keystone species: Measures of positional importance in food webs. *Oikos, 112*, 535–546.

Jordán, F., Liu, W. C., & Mike, Á. (2009). Trophic field overlap: A new approach to quantify keystone species. *Ecological Modelling, 220*, 2899–2907.

Jordán, F., Liu, W. C., & van Veen, F. J. F. (2003). Quantifying the importance of species and their interactions in a host-parasitoid community. *Community Ecology, 4*, 79–88.

Jordán, F., & Scheuring, I. (2002). Searching for keystone in ecological networks. *Oikos, 99*, 607–612.

Jordán, F., Takacs-Santa, A., & Molnar, I. (1999). A reliability theoretical quest for keystones. *Oikos, 86*, 453–462.

Jordano, P. (2016). Chasing ecological interactions. *PLoS Biology, 14*(9).

Jørgensen, P. S., Folke, C., & Carroll, S. P. (2019). Evolution in the anthropocene: Informing governance and policy. *Annual Review of Ecology, Evolution, and Systematics, 50*, 527–546.

Kondoh, M. (2003). Foraging adaptation and the relationship between food-web complexity and stability. *Science, 299*, 1388–1391.

Kortsch, S., Primicerio, R., Fossheim, M., Dolgov, A. V., & Aschan, M. (2015). Climate change alters the structure of arctic marine food webs due to poleward shifts of boreal generalists. *Proceedings of the Royal Society of London - Series B: Biological Sciences, 282*(1814), 20151546.

Krause, A. E., Frank, K. A., Mason, D. M., Ulanowicz, R. E., & Taylor, W. W. (2003). Compartments revealed in food-web structure. *Nature, 426*, 282–285.

Laigle, I., Aubin, I., Digel, C., Brose, U., Boulangeat, I., & Gravel, D. (2018). Species traits as drivers of food web structure. *Oikos, 127*(2), 316–326.

Lau, M. K., Borrett, S. R., Baiser, B., Gotelli, N. J., & Ellison, A. M. (2017). Ecological network metrics: Opportunities for synthesis. *Ecosphere, 8*(8), e01900.

Libralato, S., Christensen, V., & Pauly, D. (2006). A method for identifying keystone species in food web models. *Ecological Modelling, 195*, 153–171.

Litzow, M. A., & Urban, D. (2009). Fishing through (and up) Alaskan food webs. *Canadian Journal of Fisheries and Aquatic Sciences, 66*(2), 201–211.

Loeuille, N. (2010). Consequences of adaptive foraging in diverse communities. *Functional Ecology, 24*, 18–27.

Lotze, H. K., Coll, M., & Dunne, J. A. (2011). Historical changes in marine resources, food web structure and ecosystem functioning in the Adriatic Sea, Mediterranean. *Ecosystems, 14*(2), 198–222.

Lotze, H. K., & Worm, B. (2009). Historical baselines for large marine animals. *Trends in Ecology & Evolution, 24*(5), 54–262.

Luczkovich, J. J., Borgatti, S. P., Johnson, J. C., & Everett, M. G. (2003). Defining and measuring trophic role similarity in food webs using regular equivalence. *Journal of Theoretical Biology, 220*(3), 303–321.

Marina, T. I., Saravia, L. A., Cordone, G., Salinas, V., Doyle, S. R., & Momo, F. R. (2018). Architecture of marine food webs: To be or not be a 'small-world'. *PLoS One, 13*(5), e0198217.

Martinez, N. D. (1992). Constant connectance in community food webs. *The American Naturalist, 139*, 1208–1218.

Maschner, H. D., Betts, M. W., Cornell, J., Dunne, J. A., Finney, B., Huntly, N., et al. (2009). An introduction to the biocomplexity of Sanak Island, western Gulf of Alaska 1. *Pacific Science, 63*(4), 673–710.

May, R. M. (1972). Will a large complex system be stable? *Nature, 238*(5364), 413–414.

May, R. M. (1973). Qualitative stability in model ecosystems. *Ecology, 54*(3), 638–641.

May, R. M., Beddington, J. R., Clark, C. W., Holt, S. J., & Laws, R. M. (1979). Management of multi-species fisheries. *Science, 205*, 267–277.

McCauley, D. J., Pinsky, M. L., Palumbi, S. R., Estes, J. A., Joyce, F. H., & Warner, R. R. (2015). Marine defaunation: Animal loss in the global ocean. *Science, 347*(6219), 1255641.

McMahon, S. M., Miller, K. H., & Drake, J. (2001). Networking tips for social scientists and ecologists. *Science, 293*, 1604–1605.

Menge, B. A. (1995). Indirect effects in marine rocky intertidal interaction webs: Patterns and importance. *Ecological Monographs, 65*, 21–74.

Montoya, J. M., Pimm, S. L., & Solé, R. V. (2006). Ecological networks and their fragility. *Nature, 442*(7100), 259–264.

Montoya, J. M., & Solé, R. V. (2002). Small world patterns in food webs. *Journal of Theoretical Biology, 214*(3), 405–412.

Myers, R. A., Baum, J. K., Shepherd, T. D., Powers, S. P., & Peterson, C. H. (2007). Cascading effects of the loss of apex predatory sharks from a coastal ocean. *Science, 315*, 1846–1850.

Myers, R. A., Hutchings, J. A., & Barrowman, N. J. (1996). Hypothesis for the decline of cod in the North Atlantic. *Marine Ecology Progress Series, 68*, 293–308.

Navia, A. F. (2013). *Función ecológica de tiburones y rayas en un ecosistema costero del Pacífico colombiano* (PhD dissertation. Instituto Politécnico Nacional, Centro Interdisciplinario de Ciencias Marinas, La Paz, Baja California Sur, Mexico), 170 pp.

Navia, A. F., Cortés, E., & Cruz-Escalona, V. H. (2012). Use of network analysis in food web conservation. *Current Conservation, 6*(4), 18–21.

Navia, A. F., Cortés, E., Jordán, F., Cruz-Escalona, V. H., & Mejía-Falla, P. A. (2012). Changes to marine trophic networks caused by fishing. In *Diversity of ecosystems*. InTech.

Navia, A. F., Cortés, E., & Mejía-Falla, P. A. (2010). Topological analysis of the ecological importance of elasmobranch fishes: A food web study on the Gulf of Tortugas, Colombia. *Ecological Modelling, 221*, 2918–2926.

Navia, A. F., Cruz-Escalona, V. H., Giraldo, A., & Barausse, A. (2016). The structure of a marine tropical food web, and its implications for ecosystem-based fisheries management. *Ecological Modelling, 328*, 23–33.

Navia, A. F., Maciel-Zapata, S. R., González-Acosta, A. F., Leaf, R. T., & Cruz-Escalona, V. H. (2019). Importance of weak trophic interactions in the structure of the food web in La Paz Bay, southern Gulf of California: A topological approach. *Bulletin of Marine Science, 95*(2), 199–215.

Navia, A. F., & Mejía-Falla, P. A. (2016). Fishing effects on elasmobranchs from the Pacific Coast of Colombia. *Universitas Scientiarum, 21*(1), 9–22.

Newman, M. E. J., & Girvan, M. (2004). Finding and evaluating community structure in networks. *Physical Review E, 69*, 026113–026115.

Nixon, S. W. (1995). Coastal marine eutrophication: A definition, social causes, and future concerns. *Ophelia, 41*, 199–219.

Olesen, J. M., Bascompte, J., Dupont, Y. L., & Jordano, P. (2007). The modularity of pollination networks. *Proceedings of the National Academy of Sciences of the United States of America, 104*, 19891–19896.

Olmo-Gilabert, R., Navia, A. F., de la Cruz-Agüero, G., Molinero, J. C., Sommer, U., & Scotti, M. (2019). Body size and mobility explain species centralities in the Gulf of California food web. *Community Ecology, 20*(2), 149–160.

Ortiz, M., Campos, L., Berrios, F., Rodriguez-Zaragoza, F., Hermosillo-Nuñez, B., & González, J. (2013). Network properties and keystoneness assessment in different intertidal communities dominated by two ecosystem engineer species (SE Pacific coast): A comparative analysis. *Ecological Modelling, 250*, 307–318.

Ortiz, M., Hermosillo-Nuñez, B., González, J., Rodríguez-Zaragoza, F., Gómez, I., & Jordán, F. (2017). Quantifying keystone species complexes: Ecosystem-based conservation management in the King George Island (Antarctic Peninsula). *Ecological Indicators, 81*, 453–460.

Ortiz, M., Rodriguez-Zaragosa, F., Hermosillo-Nunez, B., & Jordán, F. (2015). Control strategy scenarios for the alien lionfish *Pterois volitans* in Chinchorro Bank (Mexican Caribbean) based on semi-quantitative loop network analysis. *PLoS One, 10*(6), 0130261.

Oshima, M. C., & Leaf, R. T. (2018). Understanding the structure and resilience of trophic dynamics in the northern Gulf of Mexico using network analysis. *Bulletin of Marine Science, 94*(1), 21–46.

Pacheco, J. M., Traulsen, A., & Nowak, M. A. (2006). Coevolution of strategy and structure in complex networks with dynamical linking. *Physical Review Letters, 97*, 258103.

Paine, R. T. (1969). A note on trophic complexity and community stability. *The American Naturalist, 103*(929), 91–93.

Pandolfi, J. M., Bradbury, R. H., Sala, E., Hughes, T. P., Bjorndal, K. A., Cooke, R. G., et al. (2003). Global trajectories of the long-term decline of coral reef ecosystems. *Science, 301*(5635), 955–958.

Pascual, M., & Dunne, J. A. (Eds.). (2006). *Ecological networks: Linking structure to dynamics in food webs*. Oxford University Press.

Pauly, D., Christensen, V., Dalsgaard, J., Froese, R., & Torres, F. (1998). Fishing down marine food webs. *Science, 279*, 197–212.

Pauly, D., & Zeller, D. (2016). Catch reconstructions reveal that global marine fisheries catches are higher than reported and declining. *Nature Communications, 7*, 10244.

Pereira, J., & Jordán, F. (2017). Multi-node selection of patches for protecting habitat connectivity: Fragmentation versus reachability. *Ecological Indicators, 81*, 192–200.

Pereira, J., Saura, S., & Jordán, F. (2017). Single-node vs. multi-node centrality in landscape graph analysis: Key habitat patches and their protection for 20 bird species in NE Spain. *Methods in Ecology and Evolution, 8*(11), 1458–1467.

Pérez-Matus, A., Ospina-Alvarez, A., Camus, P. A., Carrasco, S. A., Fernandez, M., Gelcich, S., et al. (2017). Temperate rocky subtidal reef community reveals human impacts across the entire food web. *Marine Ecology Progress Series, 567*, 1–16.

Pimm, S. L. (1979). Structure of food webs. *Theoretical Population Biology, 16*(2), 144–158.

Pimm, S. L. (2002). *Food webs*. Chicago, IL: University of Chicago Press.

Pocock, M. J., Evans, D. M., Fontaine, C., Harvey, M., Julliard, R., McLaughlin, O., et al. (2016). The visualisation of ecological networks, and their use as a tool for engagement, advocacy and management. In *Advances in ecological research* (Vol. 54, pp. 41–85). Academic Press.

Poisot, T., Stouffer, D. B., & Gravel, D. (2015). Beyond species: Why ecological interactions vary through space and time. *Oikos, 124*, 243–251.

Poos, J. J., & Rijnsdorp, A. D. (2007). An "experiment" on effort allocation of fishing vessels: The role of interference competition and area specialization. *Canadian Journal of Fisheries and Aquatic Sciences, 64*, 304–313.

Raimundo, R. L. G., Guimarães, P. R., Jr., & Evans, D. M. (2018). Adaptive networks for restoration ecology. *Trends in Ecology & Evolution, 33*(9), 664–675.

Raimundo, R. L. G., Marquitti, F. M. D., de Andreazzi, C. S., Pires, M. M., & Guimarães, P. R. (2018). Ecology and evolution of species-rich interaction networks. In *Ecological networks in the tropics* (pp. 43–58). Cham, Switzerland: Springer.

Rezende, E. L., Albert, E. M., Fortuna, M. A., & Bascompte, J. (2009). Compartments in a marine food web associated with phylogeny, body mass, and habitat structure. *Ecology Letters, 12*(8), 779–788.

Ritchie, E. G., & Johnson, C. N. (2009). Predator interactions, mesopredator release and biodiversity conservation. *Ecology Letters, 12*(9), 982–998.

Rohr, R. P., & Bascompte, J. (2014). Components of phylogenetic signal in antagonistic and mutualistic networks. *The American Naturalist, 184*(5), 556–564.

Rohr, R. P., Saavedra, S., & Bascompte, J. (2014). On the structural stability of mutualistic systems. *Science, 345*(6195).

Saravia, L. A., Marina, T. I., De Troch, M., & Momo, F. R. (2018). Ecological network assembly: How the regional meta web influence local food webs. *BioRxiv* 340430.

Sethi, S. A., Branch, T. A., & Watson, R. (2010). Global fishery development patterns are driven by profit but not trophic level. *Proceedings of the National Academy of Sciences, 107*, 12163–12167.

Shepherd, T. D., & Myers, R. A. (2005). Direct and indirect fishery effects on small coastal elasmobranchs in the northern Gulf of Mexico. *Ecology Letters, 8*(10), 1095–1104.

Solé, R. V., & Montoya, M. (2001). Complexity and fragility in ecological networks. *Proceedings of the Royal Society of London - Series B: Biological Sciences, 268*(1480), 2039–2045.

Standström, O., Larsson, A., Andersson, J., Appelberg, M., Bignert, A., & Helene, E. K. (2005). Three decades of Swedish experience demonstrates the need for integrated long-term monitoring of fish in marine coastal areas. *The Water Quality Research Journal of Canada, 40*, 233–250.

Stevens, J. D., Bonfil, R., Dulvy, N. K., & Walker, P. A. (2000). The effects of fishing on sharks, rays, and chimeras (chondrichthyans), and the implications for marine ecosystems. *ICES Journal of Marine Science, 57*, 476–494.

Suweis, S., Simini, F., Banavar, J. R., & Maritan, A. (2013). Emergence of structural and dynamical properties of ecological mutualistic networks. *Nature, 500*(7463), 449–452.

Stouffer, D. B., & Bascompte, J. (2011). Compartmentalization increases food-web persistence. *Proceedings of the National Academy of Sciences of the United States of America, 108*, 3648–3652.

Strogatz, S. H. (2001). Exploring complex networks. *Nature, 410*(6825), 268–276.

Thébault, E., & Fontaine, C. (2010). Stability of ecological communities and the architecture of mutualistic and trophic networks. *Science, 329*, 853–856.

Tromeur, E., & Loeuille, N. (2018). Adaptive harvesting drives fishing down processes, regime shifts, and resilience changes in predator-prey systems. *BioRxiv* 290460.

Valdovinos, F. S., Ramos-Jiliberto, R., Garay-Narvaez, L., Urbani, P., & Dunne, J. A. (2010). Consequences of adaptive behaviour for the structure and dynamics of food webs. *Ecology Letters, 13*, 1546–1559.

Vázquez, D. P., Melián, C. J., Williams, N. M., Blüthgen, N., Krasnov, B. R., & Poulin, R. (2007). Species abundance and asymmetric interaction strength in ecological networks. *Oikos, 116*(7), 1120–1127.

Vinagre, C., Costa, M. J., Wood, S. A., Williams, R. J., & Dunne, J. A. (2019). Potential impacts of climate change and humans on the trophic network organization of estuarine food webs. *Marine Ecology Progress Series, 616*, 13–24.

Vitousek, P. M., D'antonio, C. M., Loope, L. L., Rejmanek, M., & Westbrooks, R. (1997). Introduced species: A significant component of human-caused global change. *New Zealand Journal of Ecology, 21*(1), 1–16.

Wasserman, S., & Faust, K. 1994. *Social network analysis: Methods and applications* (Vol. 8). Cambridge University Press.

Watts, D. J., & Strogatz, S. H. (1998). Collective dynamics of 'small-world' networks. *Nature, 393*, 440–442.

Webb, C. O., Ackerly, D., McPeek, M. A., & Donoghue, M. J. (2002). Phylogenies and community ecology. *Annual Review of Ecology and Systematics, 33*(1), 475–505.

Worm, B., Barbier, E. B., Beaumont, N., Duffy, J. E., Folke, C., Halpern, B. S. et al. (2006). Impacts on biodiversity loss on ocean ecosystem services. *Science, 314*, 787–790.

Yen, J. D., Cabral, R. B., Cantor, M., Hatton, I., Kortsch, S., Patrício, J., & Yamamichi, M. (2016). Linking structure and function in food webs: Maximization of different ecological functions generates distinct food web structures. *The Journal of Animal Ecology, 85*, 537–547.

Yodzis, P. (2000). Diffuse effects in food webs. *Ecology, 81*(1), 261–266.

Zhang, F., Hui, C., & Terblanche, J. S. (2011). An interaction switch predicts the nested architecture of mutualistic networks. *Ecology Letters, 14*, 797–803.

Modelling and Conservation of Coastal Marine Ecosystems in Latin America

Marco Ortiz and Ferenc Jordán

1 Overview and Synthesis

In order to understand and maybe predict the complex behaviour of ecological systems, we need to look for general patterns and laws. Modelling is the way how to elegantly describe natural phenomena, where mathematics is as close as possible to empirical findings. It is widely accepted that any model is an intellectual construct, not pictures of reality, we analysed instead of making experiments in the real world. During this process, we distort the ecological system under study, simplifying nature in a way that preserves the fundamental features of the problem. We choose to omit some aspects of reality, grouping variables, considering as inseparable meanwhile they are different. We decide what components will be treated as variables and other as constant and we simplify mathematical relations. Despite this, the models permit us making natural processes more understandable and manageable but also adding new sources of error. "The art of model building consist in knowing when a simplification continues to promote (*facilitate, increase*) understanding and when its usefulness to become an oversimplification that obscures more that is reveals" (Puccia & Levins, 1985, p. 9; emphasis is not in original).

M. Ortiz (✉)
Laboratorio de Modelamiento de Sistemas Ecológicos Complejos (LAMSEC), Instituto Antofagasta, Universidad de Antofagasta, Antofagasta, Chile

Instituto de Ciencias Naturales Alexander von Humboldt, Facultad de Ciencias Naturales y Recursos Biológicos, Universidad de Antofagasta, Antofagasta, Chile

Departamento de Biología Marina, Facultad de Ciencias del Mar, Universidad Católica del Norte, Coquimbo, Chile
e-mail: marco.ortiz@ucn.cl

F. Jordán
Balaton Limnological Institute, Centre for Ecological Research, Tihany, Hungary

Stazione Zoologica Anton Dohrn, Napoli, Italy

M. Ortiz, F. Jordán (eds.), *Marine Coastal Ecosystems Modelling and Conservation*, https://doi.org/10.1007/978-3-030-58211-1_10

Scientists clearly wish to build and work with useful models which permit us maximizing generality, realism, and precision; however, the symmetrical overlapping of these attributes seems to be impossible when the research goals are understanding, predicting, or maybe modifying nature. As a consequence of that, three alternative and complementary strategies for model building were proposed: (1) sacrifice generality to realism and precision, (2) sacrifice realism to generality and precision, and (3) sacrifice precision to realism and generality (Levins, 1966; Puccia & Levins, 1985). Most models in this book belong to type (1) and (2). Since several authors have debated about this trichotomy strategy for model building (Justus, 2006; Odenbaugh, 2006; Orzack & Sober, 1993; Palladino, 1990), we have to note that models do not fall into mutually exclusive types but they lie on a multidimensional continuum constituted by these three axes, including additionally manageability and understandability (Levins, 1993).

Reductionistic ecology studies the effect of species i on species j, in time x and space y, under circumstances z. Spatiotemporal studies systematically analyse the effects of x and y, comparative studies focus on z, and multi-species analyses consider several i and j pairs. Holistic ecology tries to integrate all these efforts, clearly at the expense of details. Instead of microscopes, systems ecologists use macroscopes, in the form of systems models.

Beyond the natural complexity of ecological systems, we also need to understand the complexity of socio-ecological relationships, including pollution, the effects of climate change, acidification, artisanal fisheries, industrial overfishing, oil and gas exploitation, the loss of biodiversity, species introductions. Modelling these human impacts on marine coastal ecosystems with a view towards control and protection of policies encounters at least four kinds of difficulties. First, the ecosystem we intervene is a network of interacting organisms. In consequence, any natural or anthropogenic intervention percolates through that network being buffered along some pathways, amplified along others, and may even be inverted. Second, is that our own interventions are frequently not constant, that is, we act on the system, but also respond to it; therefore, our actions are in co-variation with the variables of the natural systems. Thus, human interventions could introduce more uncertainties. Third, productive marine coastal areas exhibit a large physical, biological, and biophysical variability, confounding the putative beneficial impacts of management regulations on exploited species. The fourth difficulty is that all human interventions are being severely stressed by the effect of global climate change, which could enhance their negative impacts on marine biodiversity and ecosystem properties.

The difficulties above mentioned show that the prediction of the changes in natural systems caused by human activities is a complex task, where quotas of uncertainty will append us permanently. In this sense, we adhere to the original Laplace's statement with little changes given by Professor Levins (Levins, 2007) as follows: "*if I know the initial conditions and laws of motion 'approximately' then I can know the future of the system approximately*". Therefore, we are faced with urgent problems of complexity in a changing world. Different, not necessarily mutually exclusive, scientific strategies can be used to study, assess, and attempt to predict the transformations in natural complex systems as a consequence of human interventions.

These strategies include: (1) reducing the objects of study to their small components, assuming that the parts are more fundamental than the whole from which they were extracted and supposing that the properties of the whole are an epiphenomenon of the parts; (2) the statistical "democracy" of the factors, which assigns a relative weight and suppose that the factor that explains the greatest variance is the principal cause dissecting the processes in dependent and independent factors; (3) quantitative simulation, supported by the capacity of computers to obtain numerical solutions, requiring fairly precise measurements of the variables, parameters, and exact equations; and (4) qualitative or semi-quantitative modelling, which allows integration of variables from different disciplines that do not need precise and quantitative equations and that allow the integration of non-measurable variables (Ortiz & Levins, 2011, 2017). As a way to address the complexities related to the human interventions in coastal marine ecosystems, this book collects quantitative and semi-quantitative models representing coastal ecological systems from North/South-East Pacific and Mexican Caribbean.

Most ecosystems, but coastal marine ecosystems especially, are very sensitive to human impacts. Several perturbations affect a single or only a few species, changing their behaviour or spatial distribution (e.g. highly specific pollutions or overfishing regimes). Alternatively, some drivers of environmental change influence the whole ecosystem and cause systemic changes (e.g. the increasing temperature oscillations related to climate change). Most of the perturbations act only on a small set of species but the effects spread out from them and cascade through the food web, finally provoking responses at almost all other species. For this last reason, we need holistic thinking, network models, and systems-based conservation and management.

Different modelling tools and scenarios show the variability of challenges. Since the multiple components of ecological systems have both positive and negative effects on each other, any human impact can be favourable or unfavourable for particular species. For this reason, we can apply loop analysis, a qualitative tool for assessing the signs of responses (see Chapter 5). In most cases, ecological processes have important spatiotemporal variability (see chapter "The Humboldt Current Large Marine Ecosystem (HCLME) a Challenging Scenario for Modellers and Their Contribution for the Manager", calling for spatially (Ecospace, see chapter "Macroscopic Network Properties and Spatially Explicit Dynamic Model of the Banco Chinchorro Biosphere Reserve Coral Reef (Caribbean Sea) for the Assessment of Harvest Scenarios") or temporally (EcoSim, see chapter "Using Ecosystem Models to Evaluate Stock Recovery in Two Hake Species from Chile") explicit models. Considering the multispecies-nature of both perturbations and responses, some ecological network models address the issue of multi-node importance (e.g. see chapter "Macroscopic Properties and Keystone Species Complexes in Kelp Forest Ecosystems along the North-Central Chilean Coast"). As a general strategy to simplify and understand complexity, organisms in key network positions, supposedly playing critically important roles are to be identified (e.g. see chapter "The Use of Ecological Networks as Tools for Understanding and Conserving Marine Biodiversity"). Considering individual-level variability, one can decide to develop individual-based models, where populations need not be homogeneous (e.g. see chapter "Modelling the Northern Humboldt Current Ecosystem: From Winds to Predators"). For assessing the health of the whole ecosystem, global indicators need to be chosen, for example, *Ascendancy* (see chapter "How Much Biomass Must Remain in the Sea After Fishing to Preserve Ecosystem Functioning? The Case of the Sardine Fishery in the Gulf of California, Mexico"). Addressing the issue of changing habitats and niches for different species, MAXENT models support research (see chapter "The Use of Ecological Networks as Tools for Understanding and Conserving Marine Biodiversity"). For the integration of ecological and economic considerations, measures like the NPV can be suggested and integrated to ecosystem models (see chapter "Exploring Harvest Strategies in a Benthic Habitat in the Humboldt Current System").

Based on these models, the impacts of several human interventions were simulated in marine coastal ecosystems: the impacts of El Niño and La Niña events on the fishery of small fishes at Northern Humboldt Current Ecosystem (see chapter "Modelling the Northern Humboldt Current Ecosystem: From Winds to Predators"). A holistic *Keystone Species Complex* index (KSCs) which emerged using quantitative and semi-quantitative simulations could be a useful tool for fisheries management, biodiversity conservation and monitoring (see chapters "Macroscopic Properties and Keystone Species Complexes in Kelp Forest Ecosystems Along the North-Central Chilean Coast" and "The Use of Ecological Networks as Tools for Understanding and Conserving Marine Biodiversity"). Spatial simulations of coral reefs ecosystems assess the impacts of fishing rotation (Chapter 8) and the role of connectivity (see chapter "Ecological Modelling and Conservation on the Coasts of México"). Under thermodynamic constraints, the recovery or healthy properties of intervened ecosystems would be not sustainable in cases of maximization of total jobs and/or total income, suggesting for optimizing fishing yields the ecosystem health or biomass of exploited species should be recovered before (see chapter "Exploring Harvest Strategies in a Benthic Habitat in the Humboldt Current System"). The sustainability of any fishery could be achieved if a fishing limit is proposed (see chapter "How Much Biomass Must Remain in the Sea After Fishing to Preserve Ecosystem Functioning? The Case of the Sardine Fishery in the Gulf of California, Mexico"). The limits of management versus the inherent complexity and fragility of ecosystems were discussed on the examples of recurrent collapses (see chapter "The Humboldt Current Large Marine Ecosystem (HCLME) a Challenging Scenario for Modellers and Their Contribution for the Manager"). Finally, in several examples, the priorities of decision-making were discussed, as several objectives are in contrast with no perfect solution (see chapters "Using Ecosystem Models to Evaluate Stock Recovery in Two Hake Species from Chile" and "Exploring Harvest Strategies in a Benthic Habitat in the Humboldt Current System").

Overviewing various geographical areas from the North-East Pacific (see chapter "How Much Biomass Must Remain in the Sea After Fishing to Preserve Ecosystem Functioning? The Case of the Sardine Fishery in the Gulf of California, Mexico") to the South-East Pacific (see chapter "Exploring Harvest Strategies in a Benthic Habitat in the Humboldt Current System") and from the Mexican Caribbean (see chapter "Macroscopic Network Properties and Spatially Explicit Dynamic Model of the Banco Chinchorro Biosphere Reserve Coral Reef (Caribbean Sea) for the Assessment of Harvest Scenarios"), we can see that quite similar environmental challenges and human impacts are faced by coastal marine species. The presented approaches and modelling techniques are being continuously developed and improved in order to serve predictive, systems-based fisheries, conservation, and management. What could work in the open waters may not be useful enough along the coast. Because of oil production, artisanal fisheries, and pollution, socio-ecological models must integrate ecological and socio-economic processes. This is the right way how to approach reality, as humans are not only one of the many players in these systems anymore, fortunately or unfortunately.

References

Justus, J. (2006). Loop analysis and qualitative modeling: Limitations and merits. *Biology & Philosophy, 21*, 647–666.

Levins, R. (1966). The strategy of model building in population biology. *American Scientist, 54*, 421–431.

Levins, R. (1993). A response to Orzack and Sober: Formal analysis and the fluidity of science. *Quarterly Review of Biology, 68*, 547–555.

Levins, R. (2007). The butterfly ex machina. In R. Lewontin & R. Levins (Eds.), *Biology under the influence: Dialectical essays on ecology, agriculture, and health* (pp. 167–182). New York: Monthly Review Press.

Odenbaugh, J. (2006). The strategy of "The strategy of model building in population biology". *Biology & Philosophy, 21*, 607–621.

Ortiz, M., & Levins, R. (2011). Re-stocking practices and illegal fishing in northern Chile (SE Pacific coast): A study case. *Oikos, 120*, 1402–1412.

Ortiz, M., & Levins, R. (2017). Self-feedbacks determine the sustainability of human interventions in eco-social complex systems: Impacts in biodiversity and ecosystem health. *PLoS ONE, 12*(4), e017613. https://doi.org/10.1371/journal.pone.0176163.

Orzack, S., & Sober, E. (1993). A critical assessment of Levins´s The strategy of model building in population biology (1966). *Quarterly Review of Biology, 68*, 533–546.

Palladino, P. (1990). Defining ecology: Ecological theories, mathematical models, and applied biology in the 1950s and 1960s. *Journal of the History of Biology, 24*, 223–243.

Puccia, C., & Levins, R. (1985). *Qualitative modeling of complex systems: An introduction to loop analysis and time averaging* (p. 259). Cambridge: Harvard University Press.

Printed in the United States
by Baker & Taylor Publisher Services